DEVELOPING NORMS FOR THE PROVISION OF BIOLOGICAL LABORATORIES IN LOW-RESOURCE CONTEXTS

PROCEEDINGS OF A WORKSHOP

Frances E. Sharples and Micah D. Lowenthal, Rapporteurs

Policy and Global Affairs

Board on Life Sciences
Division on Earth and Life Studies

The National Academies of
SCIENCES · ENGINEERING · MEDICINE

THE NATIONAL ACADEMIES PRESS
Washington, DC
www.nap.edu

THE NATIONAL ACADEMIES PRESS 500 Fifth Street, NW Washington, DC 20001

This activity was supported by contract/grant S-ISNCT-16-CA-1035 between the National Academy of Sciences and the U.S. Department of State. Any opinions, findings, conclusions, or recommendations expressed in this publication do not necessarily reflect the views of any organization or agency that provided support for the project.

International Standard Book Number-13: 978-0-309-48653-8
International Standard Book Number-10: 0-309-48653-X
Digital Object Identifier: https://doi.org/10.17226/25311

Additional copies of this publication are available for sale from the National Academies Press, 500 Fifth Street, NW, Keck 360, Washington, DC 20001; (800) 624-6242 or (202) 334-3313; http://www.nap.edu.

Printed in the United States of America

Suggested citation: National Academies of Sciences, Engineering, and Medicine. 2019. *Developing Norms for the Provision of Laboratories in Low-Resource Contexts: Proceedings of a Workshop*. Washington, DC: The National Academies Press. doi: https://doi.org/10.17226/25311.

The National Academies of

SCIENCES • ENGINEERING • MEDICINE

The **National Academy of Sciences** was established in 1863 by an Act of Congress, signed by President Lincoln, as a private, nongovernmental institution to advise the nation on issues related to science and technology. Members are elected by their peers for outstanding contributions to research. Dr. Marcia McNutt is president.

The **National Academy of Engineering** was established in 1964 under the charter of the National Academy of Sciences to bring the practices of engineering to advising the nation. Members are elected by their peers for extraordinary contributions to engineering. Dr. C. D. Mote, Jr., is president.

The **National Academy of Medicine** (formerly the Institute of Medicine) was established in 1970 under the charter of the National Academy of Sciences to advise the nation on medical and health issues. Members are elected by their peers for distinguished contributions to medicine and health. Dr. Victor J. Dzau is president.

The three Academies work together as the **National Academies of Sciences, Engineering, and Medicine** to provide independent, objective analysis and advice to the nation and conduct other activities to solve complex problems and inform public policy decisions. The National Academies also encourage education and research, recognize outstanding contributions to knowledge, and increase public understanding in matters of science, engineering, and medicine.

Learn more about the National Academies of Sciences, Engineering, and Medicine at **www.national-academies.org**.

PLANNING COMMITTEE ON DEVELOPING NORMS FOR THE PROVISION OF LABORATORIES IN LOW-RESOURCE CONTEXTS

Ann M. Arvin (*Chair*), Vice Provost and Dean of Research, Stanford University
Charles Chiu, Associate Professor, University of California, San Francisco
Nancy D. Connell, Senior Scholar, Johns Hopkins Center for Health Security, Bloomberg School of Public Health
David R. Franz, Independent Consultant
Thomas G. Ksiazek, Professor and Director, High Containment Operations, Departments of Pathology and Microbiology & Immunology, University of Texas Medical Branch, Galveston National Laboratory

Staff

Micah Lowenthal, Policy and Global Affairs
Frances Sharples, Board on Life Sciences
La Tasha Morgan, Policy and Global Affairs (until March 2018)
Aanika Senn, Board on Life Sciences (until July 2018)
Hope Hare, Policy and Global Affairs (from July 2018)

Consultant

Barbara Johnson, Biosafety Biosecurity International

ACKNOWLEDGMENT OF REVIEWERS

This Proceedings of a Workshop was reviewed in draft form by individuals chosen for their diverse perspectives and technical expertise. The purpose of this independent review is to provide candid and critical comments that will assist the National Academies of Sciences, Engineering, and Medicine in making each published proceedings as sound as possible and to ensure that it meets the institutional standards for quality, objectivity, evidence, and responsiveness to the charge. The review comments and draft manuscript remain confidential to protect the integrity of the process.

We thank the following individuals for their review of this proceedings: François-Xavier Babin, Mérieux Foundation; Giovanni Cattoli, International Atomic Energy Agency; Mitsuo Isono, Japan International Cooperation Agency; and Ken Ugwu, Public Health Agency, Canada.

Although the reviewers listed above provided many constructive comments and suggestions, they were not asked to endorse the content of the proceedings nor did they see the final draft before its release. The review of this proceedings was overseen by Jerry Keusch, Boston University. He was responsible for making certain that an independent examination of this proceedings was carried out in accordance with the standards of the National Academies and that all review comments were carefully considered. Responsibility for the final content rests entirely with the rapporteurs and the National Academies.

CONTENTS

ACRONYMS

BEP	U.S. Department of State's Biosecurity Engagement Program
BSL	biological safety level
BSAT	Biological Select Agents and Toxins
BWC	Biological Weapons and Toxins Convention
Cas	CRISPR-associated system
CDC	U.S. Centers for Disease Control and Prevention
CLIA	Clinical Laboratory Improvement Amendments
CLSI	Clinical and Laboratory Standards Institute
CRISPR	Clustered Regularly Interspaced Short Palindromic Repeats
DNA	deoxyribonucleic acid
ELISA	enzyme-linked immunosorbent assay
ESI	electron spray ionization
FAO	United Nations' Food and Agriculture Organization
FDA	U.S. Food and Drug Administration
HEPA	high-efficiency particulate air
HIV	human immunodeficiency virus
IHR	International Health Regulations
IU	international units
LBM	WHO's Laboratory Biosafety Manual
MALDI	Matrix Assisted Laser Desorption/Ionization
MS	mass spectrometry
OIE	World Organisation for Animal Health
PCR	polymerase chain reaction
PMF	peptide mass fingerprint
PPP	Potential Pandemic Pathogen
PVS	Performance of Veterinary Services
RNA	ribonucleic acid
SARS	Severe Acute Respiratory Syndrome
SURPI	Sequence-based Ultra-Rapid Pathogen Identification

TB	tuberculosis
TOF	time of flight
VHF	viral hemorrhagic fever
WHO	World Health Organization

OVERVIEW

On June 27-28, 2018, the U.S. National Academies of Sciences, Engineering, and Medicine (the National Academies) convened an international workshop in Amsterdam, the Netherlands, on developing norms for the provision of laboratories in low-resource contexts. The U.S. Department of State's Biosecurity Engagement Program requested that the National Academies organize this workshop to engage an international group of organizations that provide funding for construction, upgrades, and maintenance of biological laboratories in countries without the means to build such labs themselves. Twenty-one people from 19 organizations participated. The intent was to advance the conversation about the identification and application of guiding principles and common norms for use by these organizations in their grants, partnerships, and aid.

Several observations made by participants were highlighted at the workshop and are repeated here. Inclusion of an observation does not imply a consensus view of the workshop participants or the planning committee.

1. The community of funders for biological laboratories in partner countries includes development and security agencies of national governments, international organizations, development banks, scientific and clinical health organizations, and foundations. Other stakeholders include the recipient countries, regional health organizations, academic institutions, and private industry, including companies that provide advice, equipment, construction services, and supplies. It is very unusual for representatives of these groups to meet all together.
2. Different funders have different models of assistance and partnership, ranging from limited-duration projects aimed at identifying needs, constructing facilities, and training personnel, to open-ended partnerships with recipients that lead to committed collaborations over decades.

3. According to several participants, there is interest within the funder community to share best practices and information to improve outcomes for all involved. There are other efforts to address difficulties associated with providing biological laboratories in low-resource contexts. For example, the government of Canada is working with the Chatham House, the World Organisation for Animal Health (OIE), and other organizations to develop decision tools and engineering options.
4. There is no comprehensive list of existing laboratory resources in low-resource countries. Indeed, such a list may be both impossible to compile and of limited value because of the differing laboratory purposes and local context (e.g., endemicity of disease). However, the lack of even a partial list of past or ongoing projects impedes the ability of funders to allocate resources without duplications or gaps or to foster the establishment of effective lab networks. (A partial list was developed for this workshop and is presented in Appendix E. Much of the information about laboratories is either unavailable or difficult to source. In addition, some sources of information were probably not found for the compilation of Appendix E.) Some participants indicated that they would benefit from having mechanisms to share plans and coordinate with other funding organizations.
5. Biological containment laboratories pose some safety and security risks. Context matters when assessing risk. Participants suggested several contextual factors: the lab's purpose (routine clinical diagnostics, disease surveillance, maintenance of reference samples, research, outbreak response); the degree of hazard of the pathogens being handled; whether those pathogens are endemic (containment requirements may differ if the pathogen is already present in the local environment); lab personnel adherence to safety and security protocols; the regularity and effectiveness of inspections; and the adequacy and reliability of funding, electricity, water, waste treatment, transportation, supply chains, and internet and telecommunications. For a partnership to be successful, the stakeholders must work together before the project's start to align the purpose of the planned laboratory with the needs and capabilities of the host or recipient. This has been a central focus of the Canadian government's work with Chatham House and OIE.
6. A capable workforce and a strong training program are essential to the proper functioning of a biological laboratory. Many participants

suggested that funders should address knowledge gaps in regions with inadequate educational systems and should arrange for provision of training by local sources (preferable, if possible), third parties (professional societies, biosafety organizations, and private companies), or their own personnel (if a technical organization). They also stressed the importance of leadership skills, career planning, and promotion opportunities to retain workers. Funders could also engage with local or foreign universities to provide the appropriate education and training.

7. New molecular techniques that allow for work with inactivated pathogens are already in use for some purposes in low-resource settings. Many of the new diagnostic methods are not, however, mature, standardized, and inexpensive enough to replace work with live pathogens using classical microbiological approaches for most purposes. (See Box 4.1.) Polymerase chain reaction (PCR)–based technology, gene sequencing, and matrix-assisted laser desorption/ionization time-of-flight (MALDI-TOF) mass spectrometry are considered well established in clinical laboratory medicine and can be cost effective, especially if external funders provide support for the necessary equipment and reagents. However, there have been numerous instances of inter-lab discrepancies in PCR results, indicating that there is performance variation among labs, perhaps due to lack of standardization. Several participants identified molecular diagnostics as increasingly important tools with the potential to increase safety and speed compared to classical techniques, and they encouraged efforts to make them more appropriate for use in low-resource settings.
8. Repurposing or upgrading existing laboratories is sometimes preferable to building new ones, some participants noted. Phased approaches that increase lab capability as well as operating can be appropriate, although it is technically difficult to renovate a lab designed for a lower biosafety level (BSL-1 or BSL-2) into a high-containment (BSL-3) facility.
9. The One Health concept is based on the fact that human and animal health are both important and intertwined. Although the principles behind this concept are sound, several participants noted that developing countries tend to be more interested in public health and that with scarce resources the agricultural sector does not receive nearly the same level of funding and attention.
10. Effective implementation of the International Health Regulations (IHR) requires that all 196 WHO member countries have adequate

legal and regulatory frameworks. However, many low-resource countries lack not only such frameworks but also formal requirements for the operation and management of biocontainment labs. Some participants emphasized the need for potential funders to consider the status of a country's implementation of the IHR in their funding decisions. International organizations could provide guidance to countries that need laboratory support but lack the applicable legal and regulatory frameworks to ensure safe and effective lab operation and maintenance.

11. Biosecurity has not received proper attention in low-resource countries. It is, nevertheless, just as crucial an element in the operation and maintenance of biological labs as is biosafety. Funders should ensure that recipients recognize the importance of biosecurity and biosafety and have plans to adopt and implement all required measures.

1

INTRODUCTION

On June 27-28, 2018, the U.S. National Academies of Sciences, Engineering, and Medicine (the National Academies) convened an international workshop in Amsterdam, the Netherlands, on developing norms for provision of laboratories in low-resource contexts, at the request of the U.S. Department of State's Biosecurity Engagement Program (BEP). The purpose of the workshop was to engage an international group of organizations that provide funding for construction, upgrades, and maintenance of biological laboratories in countries that could not otherwise afford them to discuss what considerations should factor into decisions whether to pursue particular partnerships and projects. The intent was to advance the conversation about the identification and application of guiding principles and common norms for use by these organizations in supporting laboratories that work with pathogens that are particularly dangerous and require special handling, equipment, and facilities—biological safety level "2 plus" and above-for public and animal health. The statement of task is presented in Appendix A. The agenda is in Appendix B. The biographical sketches for the 21 participants from 19 organization is in Appendix C.

This proceedings has been prepared by the workshop rapporteurs as a factual summary of what occurred at the workshop. The planning committee's role was limited to planning and convening the workshop. The views contained in the proceedings are those of individual workshop participants and do not necessarily represent the views of all workshop participants, the planning committee, or the National Academies.

Dr. Ann Arvin, chair of the planning committee, made welcoming remarks and explained that the participants were convened to advance the discussion on best practices toward a common approach and perhaps even common criteria or "norms" for deciding whether to provide support for specific laboratories.

Why is this discussion necessary? For some time, funders have provided this kind of assistance to developing countries with mixed results. Laboratories have been built in localities with insufficient need for them, limited funds for continued operation and maintenance, or inadequate infrastructure and availability of supplies. Lacking sufficient financial support, the laboratories have become degraded, disused, and unsafe and unsecure. In addition, some laboratories may have received financial support but lacked the ability to train personnel, again leading to problems with safety, quality, and security. These problems are not unique to laboratories in low-income countries. Laboratories have been built in the United States without sufficient consideration of the true need for them or appreciation of the expense and effort needed to sustain them. However, other challenges and risks arise in developing countries. In addition, the number of such degraded or inoperable laboratories appears to be rising, although there has not be a proper accounting of the laboratories built and by whom.

In 2011, the Academies convened an international workshop in Istanbul, Turkey, that focused on the safety and security challenges associated with the spread of high-containment laboratories. Dr. Arvin noted that multiple efforts have made progress to improve the situation. For example, the U.S. State Department issued a white paper for the 2015 Biological Weapons and Toxins Convention (BWC) Meeting of Experts that described the U.S. policy for guiding decisions on whether to fund such laboratories. Canada, which in 2018 had the chair of the G7 and the Global Partnership Against the Spread of Weapons and Materials of Mass Destruction (see www.gpwmd.com), has worked with Chatham House, the World Organisation for Animal Health (OIE), and others to develop decision tools and engineering options applicable to such laboratory projects. This has resulted in Chatham House's Sustainable Laboratories Initiative and the OIE Consultation on Sustainable Laboratories, insights from which were shared with the workshop participants by the respective leads of the two activities. Although Chatham House and OIE follow slightly different approaches, their findings have many common elements to draw on and discuss.

Dr. Arvin's introductory remarks were followed by a summary of the 2011 workshop and presentations and discussions about funder goals; the need for laboratories; practical considerations for conducting the work; efforts to update guidance on laboratory safety and security; data on what we know about the location and status of existing labs; and new technological developments to replace live culture work. Because the main workshop goal was to foster productive discussions among the participants, discussions were emphasized over presentations. Throughout this proceedings, attribution is provided only for formal presentations.

HISTORICAL OVERVIEW

Dr. Frances Sharples, Director of the National Academies' Board on Life Sciences, presented an overview of the July 2011 workshop in Istanbul, Turkey (hereafter the 2011 workshop), which was also held at the request of the State Department's BEP. The 2011 workshop in some ways amounted to a precursor to the current workshop. Many of the issues addressed in 2011 were the same as those to be discussed from the funding perspective in the 2018 workshop.

On July 10-13, 2011, 68 participants from 32 countries gathered in Istanbul for the workshop on Anticipating Biosecurity Challenges of the Global Expansion of High-Containment Biological Laboratories.[1] The meeting workshop held in partnership with the Turkish Academy of Sciences. The participants included laboratory directors, scientists, engineers, and members of governmental and nongovernmental organizations. They were active in the fields of biosafety, biosecurity, scientific research, disease surveillance, and public health. Some were from countries with a long history of operating multiple laboratories, while others were from countries that had only recently opened their first biological containment lab. Many were affiliated with groups contemplating the construction of new laboratories or interested in improving their existing facilities.

The workshop examined biosafety and biosecurity issues related to the design, construction, maintenance, and operation of high-containment biological laboratories equivalent to the U.S. Centers for Disease Control and Prevention's high-containment (BSL-3+) laboratories. Although these laboratories enable the characterization of highly dangerous human and animal pathogens, assist in disease surveillance, and serve as resources for the production of vaccines, they are complex facilities and building and operating them entails some risks.

During the course of the meeting, Sharples reported that the 2011 workshop participants discussed many aspects of the workshop topic, including:

---

[1] The report summarizing the results of the 2011 workshop is titled *Biosecurity Challenges of the Global Expansion of High-Containment Biological Laboratories: Summary of a Workshop* and is available on the National Academies Press website at https://www.nap.edu/catalog/13315.

- Technological options to meet diagnostic, research, and other goals;
- Laboratory construction and commissioning;
- Operational maintenance to provide sustainable capabilities, safety, and security; and
- Measures for encouraging a culture of responsible conduct.

To develop a sense of the current status of biocontainment laboratories around the world, participants described the history and challenges they face in their individual laboratories. Speakers described steps being taken to improve safety and security, from running training programs to implementing a variety of personnel reliability measures. Many spoke about physical security, access controls, and monitoring of pathogen inventories. The 2011 workshop participants also identified tensions in the field and suggested possible remedies.

DETERMINING AND ADOPTING APPROPRIATE SAFEGUARDS

According to Sharples, the 2011 workshop participants cited many examples of inadequate (in their opinion) implementation of biosafety and biosecurity precautions. For example, in some labs, poorly trained workers were performing aerosol-generating procedures without the benefit of personal protective equipment. Many biosafety cabinets were neither functional nor regularly inspected. For some labs, the availability of electricity and water was severely limited. Many countries have few or no regulations covering biocontainment laboratories and little or no government involvement in their operation and maintenance. In contrast, laboratories in some countries invest in cutting-edge air-handling systems and adhere to the standards for all biosafety levels regardless of the mission or setting.

On the topic of safeguards, participants made the following observations and suggestions:

- Failure to implement and use all possible safety and security measures could be considered irresponsible because it imposes the risk of accidental or deliberate pathogen release on surrounding communities.
- Regulations have not kept pace with evolving practices and engineering options, and the options available in were unnecessarily expensive and did not provide maximum risk reduction.

- There is a need to fund more applied biosafety research.
- Because of limited resources and competing funding priorities, laboratories and the communities in which they reside should define an acceptable level of risk and then select their precautions accordingly using a qualitative and/or quantitative risk analysis.

Sample and Strain Transport

The 2011 workshop participants discussed the need to balance the risks and intellectual property concerns of transporting strains and diagnostic samples to regional or other analytical facilities with the costs and risks of constructing, operating, and maintaining additional biocontainment labs and pathogen collections in situ. Numerous participants expressed frustration with what they perceived to be unnecessarily restrictive transport, import, and export regulations. They also complained about burdensome paperwork, precautions that were disproportionate to the risk, long delays in obtaining transport permission, and the need for approval by multiple officials, any of whom could block a transfer. To ameliorate some of these problems, several participants suggested continuing to engage the International Air Transport Association, the United Nations Committee of Experts on the Transport of Dangerous Goods, and national governments in dialog to better define the requirements for safe transport and to accurately characterize the associated risks.

National Regulations

Many 2011 workshop participants expressed the need for regulatory frameworks that support safe and secure research without adding undue burdens. Participants said that some laboratories work under limited or poorly enforced national regulatory frameworks, while others must comply with multiple sets of regulations to satisfy donor and national government requirements. Similarly, a lack of national and international guidance and accreditation standards is a source of frustration for laboratories seeking formal accreditation or certification. Most participants believed that implementing national regulatory frameworks and certification procedures was largely the responsibility of individual countries, but agreed that international assistance could facilitate the process. Others urged donors to simplify their regulatory requirements.

Laboratory Planning

Numerous discussions in the 2011 workshop emphasized the importance of the planning (needs assessment) phase that precedes facility design and construction or upgrade. Many participants stressed the benefits of involving all stakeholders (the local community, architects, lab directors, scientists, regulators, designers, contractors, and certifiers) in this phase as well as the design and construction phase. Some participants suggested that the planning phase consider provisions for surge capacity (i.e., temporary increases in analytical capacity in response to disease outbreaks; ways in which a new laboratory might expand and complement existing national and regional capabilities; and how emerging technologies, such as molecular diagnostics, could affect containment requirements and reduce the need for labs with the highest containment levels). Preparations for long-term sustainability, including ensuring the availability of operation and maintenance funds, equipment and reagents, and adequately trained workers with the appropriate expertise (e.g., engineers, technicians, biosafety professionals, craftspeople, and lab workers) can also begin during the planning phase.

Dr. Sharples concluded her remarks by noting that many participants identified the BWC review conference in December 2011 and the subsequent annual Experts Meetings, the International Health Regulations update in 2014, and the next revision of the World Health Organization's Laboratory Biosafety Manual[2] as opportunities for the biosafety, biosecurity, and public health communities to make changes. Biosafety associations could also assist by providing neutral national and/or regional platforms for discussions among stakeholders from multiple agencies and encouraging the adoption of a biosafety culture. Additionally, individual "champions" could take up the cause and spread the message in their countries and regions.

---

[2] Kazunobu Kojima of the World Health Organization presented on this revision during the Amsterdam workshop in 2018 (see Chapter 2), which was nearly complete at that time.

2

THE NEED FOR CONTAINMENT LABORATORIES

NEEDS AND PURPOSES FOR BIOLOGICAL LABORATORIES IN LOW-RESOURCE SETTINGS

Throughout the workshop, several speakers addressed why biological containment laboratory capacity is needed. These include safe handling of potential pandemic pathogens (PPPs) and other infectious agents in research and medicine; making accurate and rapid diagnoses for PPPs and other infectious agents to ensure appropriate medical care; detecting antimicrobial resistance in infectious agents; facilitating epidemiological investigations of infectious disease outbreaks; and detecting biological attack agents. During his presentation on the need for containment laboratories, Tom Ksiazek, DMV, used the above list as a guide, reframing the needs as follows:

- Diagnostics—identifying etiology
- Supporting clinical care—identifying cases
- Supporting epidemiology
  - identifying cases
  - identifying chains of transmission using modern molecular tools
- Supporting and carrying out ecological investigations
- Applied research
  - Testing and evaluating therapeutics—preclinical studies
  - Testing and evaluating vaccines
- Basic research—developing pathways and targets for therapeutics and vaccines

Dr. Ksiazek, who has extensive experience working in settings ranging from rudimentary field laboratories to state-of-the-art research, focused on alignment of the containment level with the laboratory purpose and the issues that funders must address when considering a laboratory project.

Basic and applied research are better and more safely performed in traditional high-containment labs in higher resource environments. But there remains a need to provide public health services and medical treatment, such as diagnostics and epidemiological work, in low-resource settings. This fact creates a dilemma for policy makers who must balance the need to improve public health with the need to prevent the use of hazardous pathogens as weapons of mass destruction. In the United States, only high-containment laboratories can conduct research with Biological Select Agents and Toxins (BSAT). These laboratories devote almost one-third of their budgets to meet the associated security requirements, which contributes to the extreme expense of doing such research. He noted, however, that the practicalities are that simpler labs in low-resource settings are capable of handling clinical and environmental samples for public health.

Diagnostic testing regimes can be carried out safely in "first tier" (non-BSL-3 and -4) facilities by trained and experienced personnel. Dr. Ksiazek stated that first tier labs usually do not cache or work with agents in quantities beyond what they receive in clinical or environmental samples and usually inactivate the materials they receive. This theme emerged throughout the workshop: that is, the importance of identifying the purpose of the laboratory when opening discussions about containment and safety and security. Public health–motivated testing in lower level laboratories does, however, require sustained adequate support for facilities, equipment, and training to protect the workers and others.

The field of etiology, that is, the branch of science concerned with the causes and origins of diseases, is a crucial component of safeguarding public health. Dr. Ksiazek stated that the failure to recognize the cause of serious disease in low-resource settings is *partially* due to a lack of diagnostic capabilities. This is problematic, because early direction of resources is very important in the fight against serious pathogens. The International Health Regulations (IHR) required rapid recognition and identification of PPPs after the epidemic of Severe Acute Respiratory Syndrome (SARS) in 2004. Delays in recognizing the serious nature of the early stages of the West African Ebola outbreak in 2014 are often cited as the reason for the disease's uncontrolled spread by the time it was accurately identified. However, Dr. Ksiazek suggested that a lack of diagnostic facilities was not altogether to blame. Even if a laboratory exists, if it is not operated daily, then the personnel who know how to operate it may not be available in a time of crisis. In other words, the prevailing local conditions may influence a laboratory's ability to fulfill its purpose.

National or regional facilities can also support rapid recognition. In-depth comparisons of outbreak agents with reference strains, both by molecular and classical techniques, are highly desirable. Therefore, the early engagement of reference labs with appropriate containment capacity is desirable. However, moving samples to and from reference laboratories is becoming more and more problematic and restricted by regulations and security concerns. Ksiazek stated that as a result, it is now almost impossible for reference laboratories to obtain export/import licenses to share pathogens, which hinders not only research, but also the pursuit of potential regional approaches. In addition, although organism strain genome sequences are documented, the strains themselves are often not appropriately preserved. Consequently, the development of treatment options becomes more difficult, because the sequences alone are insufficient for the necessary research on countermeasures.

In addition, shippers are reluctant to handle pathogens. Federal Express, for example, ships 6 million packages per day through Memphis, Tennessee. However, because an Army laboratory inadvertently sent live samples rather than inactivated samples of *Bacillus anthracis* (anthrax) to scores of other laboratories, the company announced in July 2015 that it will no longer ship research samples of Select Agents through its regular shipping service. This leaves only very specialized, dedicated delivery services (e.g., FedEx Custom Critical) to deliver such samples, which costs orders of magnitude more than do ordinary shipments.

Dr. Ksiazek discussed the need for risk assessment. He separated risk into personal risk and environmental risk. Personal risk pertains to whether the pathogen is infectious to humans or a hazard for laboratory workers (through punctures or aerosol exposures) and whether vaccines and/or treatment options are available. Environmental risk pertains to whether a pathogen is contagious, indigenous, aerosol infectious, or agricultural. He presented a graph that plots personal protection against environmental containment, showing how the various biological safety levels are distributed along these two metrics (Figure 2.1). "Inside containment" requires physical barriers, animal caging systems, laminar flow hoods, centrifuge carriers, and air exchange and gradients, with the primary personal barrier being positive pressure suits. "Environmental containment" at a typical BSL-4 includes HEPA filtration and constant negative pressure for air handling, decontamination of lab equipment and of waste (autoclaving), personnel and biological samples, and lab effluents. He explained that high-containment (BSL-3 and -4) laboratories are neither portable nor affordable for many low-resource settings. Most pathogens studied in high-containment labs in developed countries are not indigenous to the locations of these labs.

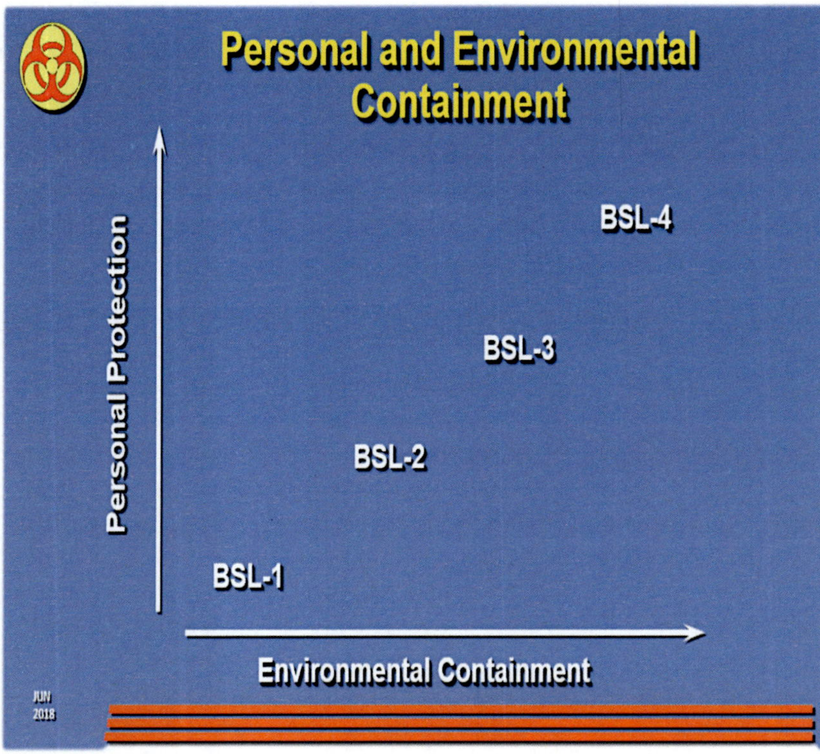

FIGURE 2.1 Personal protection plotted against environmental containment, showing the distribution of the containment levels along these two metrics. SOURCE: Thomas Ksiazek.

To conduct field studies or respond to epidemics, a researcher must travel to the environments where they exist. Dr. Ksiazek stated that there is no point in operating a BSL-4 facility for containment to prevent a pathogen from escaping to the environment if infectious patients or the organisms themselves are 20 paces away. As an example, he cited a paper by personnel from the Viral Special Pathogens Branch of the U.S. Centers for Disease Control and Prevention that described the field lab they established in Bo, Sierra Leone, during the 2014 Ebola outbreak (Flint et al., 2015). Although not a BSL-3 or BSL-4 facility, this laboratory was adequate to process more than 12,000 specimens from throughout Sierra Leone. He also cited a 2018 article (Shoemaker et al., 2018) on the impact of enhanced viral hemorrhagic fever (VHF) surveillance on outbreak detection and response in Uganda (Figure 2.2). This paper showed that surveillance leading to early detection and outbreak responses in turn led to a significant decrease in intensity and duration of VHF outbreaks in Uganda. This successful project can serve as a role model for detecting and responding to international health threats.

Dr. Ksiazek also noted that the World Health Organization (WHO) Laboratory Biosafety Manual (LBM), now under revision, will follow a risk-based approach. Diagnostic laboratories can also be used to detect antibiotic resistance using molecular techniques if the resistance mechanisms and markers are known. These techniques would probably not require high-level containment. To detect biological attack agents and to distinguish engineered agents from naturally occurring outbreaks, researchers need epidemiological information to sort outbreak situations and early sequence data to assess what species or strain the attack agent is most closely related to. Depending on the nature of the agents, such situations may require either tier 1 or regional reference labs or both.

During the discussion following Dr. Ksiazek's presentation, a participant asked whether compliance with the IHR requires BSL-3 capabilities to fulfill surveillance obligations. Dr. Ksiazek replied that a country would not need to operate a BSL-3 laboratory but would need to enter into some arrangement to access a BSL-3 laboratory when needed. He suggested that the possibility of obtaining BSL-3 lab capabilities to supplement the capabilities of existing lower-level labs might be an enticement for developing countries to comply with the IHR. Another participant stated that the risks associated with the work expected to be performed in a new or enhanced laboratory should drive the decision about what type of laboratory is really needed in that particular place. Because it may not be a BSL-3 lab, it might be preferable to move away from use of the term "BSL-3" as a descriptor. However, another participant noted that politicians in low-resource settings may ask for the highest tech facilities to bolster national pride without understanding the complexity and costs associated with such laboratories.

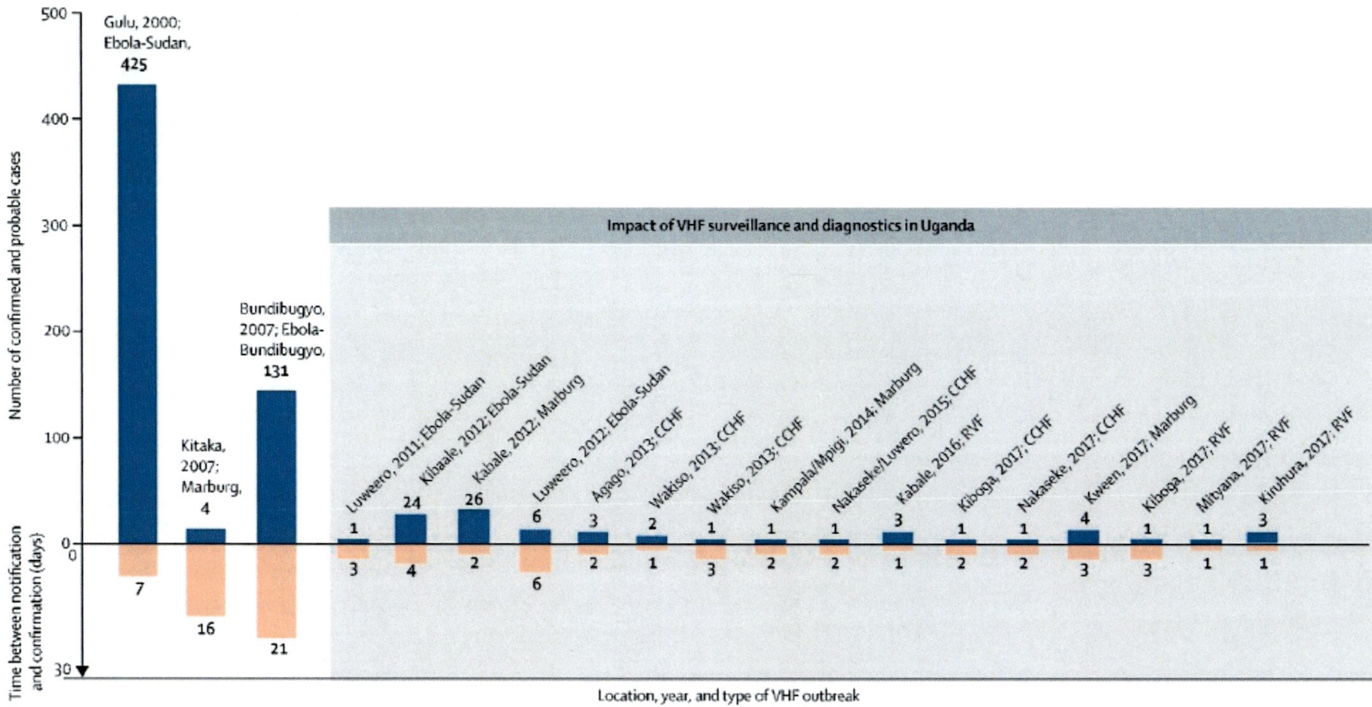

FIGURE 2.2 Surveillance leading to early detection and outbreak responses, which in turn led to a significant decrease in intensity and duration of VHF outbreaks in Uganda. SOURCE: Shoemaker et al., 2018.

REVISION OF THE WHO LABORATORY BIOSAFETY MANUAL (LBM)

Dr. Kazunobu Kojima presented on the revision of the WHO LBM, which he is leading. The current version (third edition) was published in 2004 and is outdated, given the pace of the science and technology for infectious disease. WHO member states need an up-to-date manual to guide establishment, use, and dismantlement of needed laboratories. The IHR, which are legally binding, require WHO member states to develop minimum core national and international surveillance and to report capacities. Core Capacity 8 of the IHR pertains to laboratories and calls for policy and coordination, diagnostic capacity, laboratory biosafety and biosecurity, and laboratory-based surveillance. These regulations thus oblige member nations to have certain minimum public health-related capabilities in place, which translates to some form of biological laboratory. Dr. Kojima described how this requirement may be met without establishing laboratories with high containment levels where appropriate.

Adding to Dr. Ksiazek's comments about risk assessment, Dr. Kojima noted that the LBM revision is meant to follow an "evidence-based" and "risk-based" approach, as the best way to inform the risk assessment process and policy instruments, allow logical prioritization to avoid overkill and overdesign of lab facilities, and learn from actual incidents to prevent recurrences. These considerations should facilitate resource optimization.[1] He also discussed the findings from "Surveillance of laboratory exposures to human pathogens and toxins: Canada 2016."[2] In 2016, a total of 100 lab workers were accidentally exposed with no reports of secondary exposures. Most incidents (greater than 90 percent) occurred in risk group 2 or biocontainment safety level 2 (RG2 or BSL-2) facilities. The causes were failure to follow standard operating procedures (72 percent) and equipment failure (17 percent). These statistics illustrate the importance of having, and following, biosafety requirements. Regarding

---

[1] Dr. Kojima recommended that the participants read "Evidence-Based Biosafety: A Review of the Principles and Effectiveness of Microbiological Containment Measures" by Kimman et al. (2008).

[2] Bienek, A. M. Heisz, and M. Su. 2017. "Surveillance of laboratory exposures to human pathogens and toxins: Canada 2016." *Canada Communicable Disease Report* 43-11:228-235.

risk tolerance, he asked "Who can decide what sort of capacity and authority is needed and on what grounds?"

Dr. Kojima defined "risk" as the sum of the severity of hazard plus the likelihood of exposure to the hazard. He noted that implementation of Good Microbiological Practices and Procedures (GMPP) depends on training rather than engineering controls. The Canadian data on laboratory accidents showed that the most well-designed and engineered laboratory is only as good as its least-trained worker, and that human factors are generally the cause of laboratory-acquired infections rather than malfunction of engineering controls. However, "risk (hazard) group" does not equate to biosafety level.

There are a variety of factors determining the consequences of an exposure to a pathogen, including:

- Low infectious dose
- High communicability
- Airborne route of transmission
- Availability of preventive or therapeutic treatment
- History of laboratory-acquired infection
- Exotic epidemiology (non-endemic)
- High susceptibility of population (e.g., immunocompromised, naïve)

These factors increase the severity and mortality resulting from pathogen exposure. Certain laboratory procedures increase the likelihood of exposure, as follows:

- Producing and using large volumes and high titers
- Following procedures that might generate aerosols (e.g., sonication, or deliberate generation of aerosols)
- Infecting animals
- Using sharps
- Necropsy where infection is suspected
- Increasing virulence

In contrast, some procedures present a low risk for exposures:

- Use of agar plates (e.g., streaking, spreading)
- Serial dilution
- Preparing/staining slides
- Nucleic acid extraction
- Inactivation

- Use of autoanalysers
- Enzyme-linked immunosorbent assay (ELISA)
- Polymerase chain reaction (PCR)
- Rapid diagnostic tests.

Dr. Kojima showed a graph that plots the consequences of infection against the likelihood of exposure. The revised WHO LBM will replace risk groups and biosafety levels with a thorough risk assessment and appropriate risk mitigation and control measures based on the consequence of infection from the pathogen and the risks associated with the procedures to be performed. Activities that have only "core requirements," the minimum requirements for safely executing the majority of lab procedures, because there is a low process risk and low consequence of infection. The core requirements include codes of conduct, competent and appropriately trained staff, and GMPP, and are fundamental to safe work practices in any facility.

Most work can be performed using this set of minimum core requirements. Some work will require heightened control measures (e.g., a biological safety cabinet or extra personal protective equipment, segregated work area, task-specific equipment, or a combination). Heightened control measures are required with increased risk. Very few procedures will require high containment (i.e., BSL-4) control measures. Examples include working with eradicated diseases, such as smallpox, or using procedures that entail a high likelihood of worker exposure and/or release to the environment. This risk-based approach offers much needed flexibility to the way risk is controlled.

The revision of the WHO LBM seeks to produce a central core document with additional detailed monographs on risk assessment, biosafety program management, laboratory design and maintenance, biological safety cabinets, personal protective equipment, decontamination and waste management, and emergency outbreak response. Dr. Kojima circulated a recent news article from *Science* called "Risk-based reboot for global lab biosafety,"[3] which discusses the revision of the manual that he is working on.

Finally, Dr. Kojima discussed the high heterogeneity of the regulatory situation among Member States. Some countries, such as the United

---

[3] For more information, see Kojima, K., Makison Booth, C., Summermatter, K., Bennett, A., Heisz, M., Blacksell, S.D., and McKinney, M. 2018. "Risk-based reboot for global lab biosafety: New WHO guidance could expand access to lab facilities." *Science* 360:260-262.

States, are highly regulated and have detailed biosafety and biosecurity legislation and regulations with well-defined responsibilities and processes. Other countries almost completely lack regulatory guidance in the form of legislation, standards, and regulations. WHO will undertake a new project to analyze the biosafety and biosecurity legislative framework of different WHO Member States and to develop a proposal for a harmonized international approach to ensure state-of-the art legislation for biosafety and biosecurity in biomedical laboratories.

3

THE DONOR POPULATION AND ITS GOALS IN PROVIDING BIOLOGICAL LABORATORY CAPACITY FOR LOW-RESOURCE SETTINGS

WHO FUNDS LABORATORIES IN LOW-RESOURCE SETTINGS?

As noted in Chapter 1, national governments, international organizations, development banks, public interest foundations, and private-sector entities are all funders of biological laboratories in low-resource settings. Health institutes, research agencies, academic institutions, and scientific professional societies also play roles. However, not all of the organizations that provide support for biological laboratories actually financing their construction. Some may provide equipment, reagents, and other needed supplies. Others may pay for enhancements of existing labs so that they can perform more complex and higher level work. Others may provide expert personnel to train developing country lab workers and guide research projects. All of these forms of aid are important to the sustainability and effectiveness of biological labs. The scope of the diversity of support organizations can be gleaned from the table of partial information on laboratories in Appendix E.

WHAT NEEDS ARE FUNDERS TRYING TO MEET?

To tackle this question, participants described the goals of funding organizations. For example, one country has a strong national policy for combating infectious diseases and therefore maintains programs in five African countries to provide enhanced laboratory functions, long-term training of human resources, and contributions to regional networks. This country also works in Asia and Central America. A participant stated that the donor's perspective is threefold: (1) Is there a true need for a BSL-3 facility? (2) What are the human and financial resources needed to sustain such a facility? (3) What is the likely management capacity for a particular laboratory, including biosafety factors? The "funding" modalities for this country's donations include grants, loans, and technical cooperation.

The overall goal of development banks is to reduce poverty and improve life. Banks can only provide funding, which includes to support the development of labs to improve health and fight diseases such as drug-resistant malaria and tuberculosis. Banks receive requests across the full spectrum, from aid for diagnostics to creation of national and international reference labs. Some development bank work in Africa follows the One Health approach that encompasses human and animal health. The World Bank evaluates recipient needs against the Joint External Evaluations associated with the International Health Regulations (IHR) to provide diagnostic and surveillance capabilities. To assess veterinary needs, the World Bank uses the World Organisation for Animal Health (OIE) Tool for the Evaluation of Performance of Veterinary Services (the PVS Pathway Evaluation Tool). It also tries to follow a regional approach to avoid duplication and promote efficiency through resource sharing. However, evaluation remain a big challenge, and the Bank relies on partners from OIE, the United Nations' Food and Agriculture Organization (FAO), the World Health Organization (WHO), and other outside experts for technical help. It is critical to get the project right in the planning phase.

A participant noted that funders are generally asked first to support a BSL-4 laboratory. When they refuse, they are asked to support a BSL-3 laboratory. This concerns funders, because "BSL" level is a form of branding that has led, in some cases, to a political demand rather than a need-driven demand, which can result in the proliferation of unsustainable high-containment labs that cannot maintain quality or function according to design.

Another participant stated that several funder nations' work is motivated by public health needs, seeking to strengthen health systems through capacity building. He noted that "biosecurity" is difficult to define, has a range of meanings for different people, and can be viewed from many angles.

Another participant noted that a broad set of experts from both donor and recipient countries agree that a diversity of serious problems exist with lab sustainability. A recent OIE report calls for multidisciplinary, multi-sectoral, and collaborative solutions for these issues. Here are its key conclusions:

1. A functioning, appropriately resourced laboratory contributes to prosperity, stability and security at national, regional, and global levels.

2. Laboratory facilities (including their infrastructure, engineering, and flexibility of design) must be "fit for purpose" and thus adapted to the local context and risks.
3. Sustainability of laboratory biosafety and biosecurity, quality management, and business continuity are inextricably linked.
4. Political buy-in, governance of laboratories, and empowerment of laboratory staff are key to the sustainability of a laboratory.
5. Sustainability will be improved by networking and sharing of information and best practices at all levels (local, national, regional, and international).
6. The adoption of risk-based and evidence-based approaches will make a positive contribution to sustainable laboratory biosafety and biosecurity.
7. Creative and open-minded thinking and innovation are key to improving laboratory sustainability. This includes reframing the problem, satisfying basic needs, and reasoning around the functions of the laboratory.
8. A sustainability strategy must also consider sustainable approaches to education, training, and retention of competencies.
9. Business models for sustainable laboratories and capacity building benefit from utilizing private-public-partnership models, which engage private sector users and suppliers.

The participant who referenced the OIE report believes that progress in this arena requires a "consortium approach" by multiple international organizations and donors. In particular, the report urges the human and animal health entities to work together in a coordinated way to facilitate forward progress. Another participant shared a concern about the training and mentoring of laboratory personnel, which extends beyond biosafety training to building careers, learning leadership skills, and engaging people in meaningful activities. Another participant noted the difference between training and education and added that his organization is conducting a study to assess the feasibility of creating a training hub in Africa. Although the issues for human vs. animal health are similar, the animal sector is much less well funded.

WHERE ARE THE LABS THAT HAVE ALREADY BEEN DONATED AND ARE THEY OPERATING AS ANTICIPATED?

Frances Sharples gave a brief presentation on the data compiled from several sources on the location of existing containment labs in low-resource countries.[1] The countries defined by the World Bank (2017) as Low-Income Economies and Lower Middle-Income Economies (see Appendix D) and are located in the following regions:

- Southeast Asia/Pacific
- South Asia
- Central Asia/Eastern Europe
- Middle East/North Africa
- West and Central Africa
- East Africa
- Southern Africa
- Central America and Caribbean

A table with a list of these countries organized alphabetically and the data obtained for each is located in Appendix E. The data include the number of BSL-2/3 and in one case (India) BSL-4 facilities in each country; the type of laboratory (i.e., its purpose; the funder, if known; its status (i.e., operational, nonoperational, under construction, if known); and its capabilities in terms of what tests can be performed and what categories of work (e.g., research, surveillance, sample testing), if known.

Dr. Sharples summarized the lab location information as follows:

- No single source provides a comprehensive list of lab locations and operational information.
- Many small, low-resource countries have one or a few BSL-3s and/or BSL-2s for diagnostics of region-relevant human diseases (TB, malaria, HIV), and several have one or a few BSL-3s and BSL-2s for region-relevant animal disease issues.
- Labs that possess Biological Select Agents and Toxins (BSAT) usually have pathogens relevant to diseases prevalent in their localities, such as plague (*Yersinia pestis*), Lassa fever in West

---

[1] The data were obtained from: Dr. Barbara Johnson, Biosafety Biosecurity International, and a consultant to the project, who provided information and also compiled most of the lab location data; the U.S. Defense Threat Reduction Agency's Biological Materials Information Program with the assistance of Dr. Mark Hansberger; Drs. Daniel B. Jernigan, Steve Monroe, Kevin Karem, and Inger K. Damon of the U.S. Centers for Disease Control and Prevention; and Dr. Craig Reed, CEO, Inspirion Biosciences. Much of the information on laboratories is either unavailable or difficult to source. In addition, some sources of information were probably not found for the compilation of Table E.

Africa, and Rift Valley fever in East Africa. Note that the "BSAT" designation is a U.S. classification for the most hazardous pathogens, and is not used by many other nations. The organizers do not know whether comparable designations by each of the countries on the list exist.

- Funding comes from international organizations, host and foreign governments, foundations, corporate entities, and other sources, but information was not always available about the specific funder(s).
- Some laboratories receive operation and maintenance or equipment funds from donors, although sometimes only when confronting outbreak situations.
- The list contains many unknowns and some data discrepancies. For example, a recent presentation by a director of laboratories in India provided a map of only 16 BSL-3s, while the data in the larger list indicate 44 BSL-3s in India.

Dr. Sharples concluded with the comment that a comprehensive list of laboratory location and operational data would inform the funder community's evaluation of needs and options. During the lively discussion that followed her presentation, several participants stated that they know that some of the data in the list for particular countries are inaccurate and many additional laboratories that can perform diagnostic testing are not included in the list.

David Harper started the next discussion with remarks informed by his work with Chatham House. Most recently Chatham House and Global Affairs Canada held a series of meetings that focused on improving the sustainability of laboratories built in partnerships between funding organizations and recipient countries or organizations. When he started the Sustainable Laboratories Initiative at Chatham House,[2] his first intuition was to map the environment but was quickly disabused of the notion. Playing devil's advocate, Mr. Harper asked the participants, why are we trying to map the environment? What drives our need to create such a map? In terms of norms and standards for sustainable laboratories, what will we learn from a map to drive the work forward? The answers to these questions notwithstanding, it is equally important for stakeholders to have a conversation about the energy and resources that should be devoted to mapping, given that a comprehensive map—that captures not only public

---

[2] For a description of the Sustainable Laboratories Initiative at Chatham House, see https://www.chathamhouse.org/about/structure/global-health-security/ sustainable-laboratories-initiative#.

health, but also animal and environmental health and the security sector—is probably an impossibility.

Mr. Harper questioned whether such maps really help the funding community to assess needs and options. One strand of the Chatham House work was described in a summary of a workshop focused on development of a tool to facilitate the initial dialogue between potential funding and recipient partners. In that tool, the partners are asked: Why do you need a laboratory? What work will be performed there? What alternatives exist? Who else have you approached to help build the capacity? The questions and answers may not be country specific: for example, in Nigeria, with a federal system, the questions and answers might be state specific and even city specific. The funding partner should initiate that conversation, and the recipient country needs to consider and provide that information. There needs to be some a due diligence process with appropriate verification and validation of the risks. This is challenging. At a recent meeting held in Abuja, Nigeria, participants from recipient countries said that some recipients might need help to build their expertise and capacity to do so.

An earlier speaker said he would be concerned about a local risk assessment if the local organizations do not have the expertise to do the assessment. There is a feedback loop: If the partners need a local risk assessment then it is incumbent on the people concerned, recipient and funding partner, ensure that that capacity is available in the country or locality. So returning to the point about maps, Harper argued that while it is good to have information, it should add value.

Mr. Harper noted that Chatham House partnered with the African Academy of Sciences in Nairobi to convene urban preparedness and resilience experts from Ghana, Nigeria, South Africa, and Kenya in December 2017. A group of participants who work in different but similar areas (e.g., public health military service, emergency management, veterinary service) told Mr. Harper that they had never met each other before. When studying the public health side in a nation that provides funding, he found that two offices in the same organization (not just the same government) were not liaising with each other on activities in the same country. Specifically, one office was establishing a laboratory in a country with which it was working closely on the Ebola crisis, but was unaware of directly relevant activity by another office in the same organization. The Nairobi meeting, like this Amsterdam workshop, brought together experts who should work together but almost never have the opportunity to meet. The true real merit of the current workshop was to bring together different constituencies and experts to increase awareness of others' activities and hopefully coordination.

A workshop participant stated that, because not every project can be funded, donors carefully select their investments, building toward the goal of establishing a sustainable and effective network of laboratories. It is difficult to make those decisions without knowing what already exists and who else is working in that region. It is also important to establish whether the recipient country's regulators are aware of all of the laboratories in its territory and whether they can monitor them. Beyond the desire for a map as a means to understand the status quo, these are legitimate reasons to have one, he argued. Mr. Harper questioned the definition of a "sustainable and effective network of laboratories." Is it a set of human, animal, or plant laboratories? Is it locations that store pathogens? He asserted that it is better to address specific needs than to develop a comprehensive map, even if it were possible, because the cost versus the reward is too high. Resource mapping for an individual decision may be more valuable.

Another participant pointed out that the Joint External Evaluation reports under the Global Health Security Agenda recommended creation of an overview of laboratories that work with dangerous pathogens. The Netherlands has created a pilot project database called the national inventory of dangerous pathogens, for work in Uganda. This database helps to fulfill obligations under the Biological Weapons Convention, which calls for countries to ensure the safety and security of dangerous pathogens, and which cannot be accomplished without knowing what is done and where. Another participant noted the need for donors to be aware of other funders in a country to leverage opportunities to build regional capabilities. However, a map or list of laboratory locations could serve a "dual use" in that people could use it with malicious intentions.

A participant stated that focusing on biosafety and biosecurity mapping is more important than mapping laboratories because the presence of a high- containment lab is a poor proxy for biosecurity risk. In other words, it may not be important to map the locations of particular types of labs (e.g., BSL-3s) because, in low-resource countries, dangerous pathogens are, of necessity, handled in whatever facility is available, even BSL-1 laboratories. More relevant information is where particular techniques, such as viral isolation, are practiced, and how pathogens are stored. Diagnostic labs doing polymerase chain reaction (PCR) or enzyme-linked immunosorbent assay (ELISA) are not a risk.

Another participant agreed that a map of laboratory locations would not be useful, and advocated for requiring recipients to map resources. Funders must understand the recipient's needs *and* what capabilities already exist in their country. Also needed is evaluation of the sophistication of recipient labs and their support system. The skilled-worker challenge could be addressed through professional associations as well as networks of practitioners across sectors to share best practices. However, it can be difficult for these societies to operate in some countries.

Several participants mentioned the need for skilled workers. One participant suggested asking potential recipients "What are you capable of doing?" and "Are the necessary skilled workers and equipment in place?" rather than "What do you want to do?" Another participant explained that WHO has a substantial training program and widely distributes its Laboratory Biosafety Manual (LBM). However, the LBM practices were formulated before the development of PCR, so some of its guidelines are outdated and are undergoing revision (see Chapter 2.)

A participant said that a mapping of public health assets is absolutely fundamental to inform the donor decision making process, because otherwise we accept that "more is more, and more is good," which is not always true. It is relatively easy to identify publicly funded lab assets, but it is more difficult to locate facilities provided by the private sector and academia. Resources should be allocated in the best ways possible, which entails knowing not only where the laboratories are located, but also what connections exist. Asset mapping could also be overlaid with risk mapping to seek designs based on optimal resource distribution. Such mapping would be money well spent because containment labs are extremely costly (e.g., $35 million for BSL-3 laboratories in Zambia and Ethiopia).

Another participant commented that adding to existing laboratories may be a better strategy than new construction, which should never be entered into lightly. Another participant said that the scale of the number of laboratories *below* BSL-3 is huge. For example, Pakistan believes that it has 4,000 to 6,500 laboratories, but only 2,500 are mapped. However, not all of them deal with indigenous high-hazard pathogens.

A participant agreed that the locations of laboratories should be driven by local risk assessments and needs, but a map of BSL-3 and -4 laboratories is needed from a defense/security perspective to ensure that these labs are secure and that pathogens are being handled appropriately. Another participant noted that it is important to know how long funding will remain available. A third person agreed that we need to map the techniques that are in use and the pathogens being worked with or stored. Out of necessity, researchers will perform the work at whatever level facility is available, which is a concern. Another important piece of information is the operational status of existing labs.

A participant stated that mapping the location of pathogens for security purposes is puzzling. Pathogens regularly pass through clinical and diagnostic labs wherever they are endemic, and maps of disease in broad regions are currently available. It is important to distinguish laboratories that store an isolated pathogen for research and reference from the places where it can be found in nature. Furthermore, it is not clear that a person or group with malevolent intent will use the appropriate BSL lab for the pathogen, or any lab at all.

A participant stated that evidence-based biosafety is necessary in resource-limited contexts to keep cost and complexity down. Evidence should inform the risk-assessment process and policy, allow for logical prioritization, lead to learning from past lessons, and result in optimization of resources.

Another participant asked whether donors have a preference for building new facilities rather than supporting existing facilities. At least one donor government does, but no one else in the room answered this question, perhaps because the answer depends on the particular donor or is not precisely known. One participant answered only that his government does not like to create brand new labs. It prefers to enhance or expand an existing laboratory that is deemed suitable for the purpose. Another participant agreed that expanding the capacity of existing facilities can be better than new construction, but "mapping" is still required to determine those needs. The participant who believes strongly that mapping per se is not useful underscored the need for clarity about what is being mapped. Samples of hazardous pathogens are extensive where these are endemic. What the donor wants is important, but a laboratory will only be sustainable if attention is also paid to what the recipient wants.

4

MOLECULAR DIAGNOSTICS: AN ALTERNATIVE TO HIGH CONTAINMENT?

Charles Chiu, MD, PhD, a member of the workshop planning committee, gave a presentation titled “Molecular Diagnostics in Low-Resource Settings.” He described the “classical” (i.e., non-molecular) microbiological testing methods as including:

- Culture
- Serology
- Microscopy
- Biochemical profiling
- Direct antigen testing: Lateral flow immunoassays and matrix assisted laser desorption/ionization (MALDI) for bacterial, viral, and fungal identification

Molecular diagnostics, or “DNA-Based detection,” include a variety of new, and even experimental, technologies, such as:

- Hybridization (probes), for example, clustered regularly interspersed short palindromic repeats (CRISPR)-Cas based assays
- Genotyping
- Sequencing, including nanopore sequencing
- Signal amplification
- Target amplification (polymerase chain reaction [PCR]): Single-plex and multiplex

Molecular diagnostic tests offer some advantages for low-resource settings. The pathogens are inactivated for testing, so handling them is safer than in methods that require the use of infectious live organisms, decreasing the potential for occupational exposures. They do not rely on culture-based amplification, which is important because many pathogens are not culturable. Because such molecular-based testing enables performance of diagnostics with noninfectious inactivated pathogens, the need for costly and complex BSL-3 or -4 containment is obviated for diagnostic work on very hazardous pathogens. Molecular methods also offer faster turnaround time and do not require large sample volumes.

However, Dr. Chiu continued, molecular testing also has significant disadvantages. First, these methods are more expensive than classical techniques. One participant noted, for example, that the cost of one PCR kit equals about 1 year's salary for a lab worker in low-resource settings. Second, performance assessment, validation, and regulatory approval of many of these methods are challenging, especially if the work is performed outside of highly controlled clinical laboratory environments, which may not be available in low-resource settings. Dr. Chiu reviewed some areas in which standards for molecular testing are lacking, particularly for environments that are not highly regulated: positive and negative controls, platforms, analytical performance, target pathogens, and reference databases. Because of this lack of standardization, the same assay run in two different labs may yield different results, and confirmatory testing is slow and costly.

In addition, the entities that normally certify and/or approve such tests (e.g., the U.S. Food and Drug Administration [FDA], the U.S. National Institute of Standards and Technology, the World Health Organization (WHO), and various nongovernmental organizations) have not yet done so for most of the new molecular technologies. In the United States, a regulatory framework, the Clinical Laboratory Improvement Amendments (CLIA), guides clinical laboratory testing. It sets minimum standards under which all clinical laboratories operate. CLIA laboratories are certified by inspection by an agency such as the College of American Pathology. Compliance with CLIA requires validation and quality assurance for all laboratory tests used in clinical care, including "laboratory-developed tests." The Clinical and Laboratory Standards Institute (CLSI) also issues "Guidelines—CLSI Molecular Diagnostic Methods for Infectious Diseases" (CLSI, 2015). But certification under these frameworks is not the same as FDA approval. Furthermore, many laboratories in low-resource settings may not meet CLIA or similar regulatory standards for proficiency testing, incorporation of standardized controls, etc.

Pre-analytical, analytical, and post-analytical concerns exist for molecular diagnostics, Dr. Chiu explained. Pre-analytical concerns include the need for proper sample collection methods, appropriate timing, proper storage conditions for both organisms and assay components (e.g., maintenance of a cold chain, which is especially important in low-resource settings and with labile ribonucleic acid [RNA]), and control of contamination. Analytical concerns focus on test performance evaluation—sensitivity, specificity, precision, accuracy, linearity, matrix effect, interference, reproducibility, and limitations. Post-analytical concerns include proper reporting of results, copies/ml or IU/ml or Log IU/ml, positive and negative predictive values (PPV and NPV, respectively), and diagnostic value and clinical utility.

Based on these and other considerations, Dr. Chiu provided a list of relevant questions that should be addressed in the context of molecular testing in low-resource settings:

- Who will pay for a test, and who is trained and certified to run it?
- How often will the test be run? What volumes of material are required? Do the particular circumstances justify the costs?
- Can clinically significant organisms be identified and quantitated in patient specimens or from culture?
- If culture is not possible, molecular methods may be justified. But if they are to be used for prognosis, surveillance to guide public health interventions, or diagnosis to guide therapy for individual patients, will they provide the necessary accurate information?
- Where will a test be run and in what settings? Are these settings appropriate for achieving accurate results?

Dr. Chiu then briefly described each technology category that he listed at the start of his presentation and commented on their states of development (see Box 4.1 at the end of this chapter).

Dr. Chiu explained that direct detection methods, such as sequencing, cannot fully replace serology, the branch of laboratory medicine that investigates blood serum to detect antibodies and antigens.[1] He views molecular testing as complementary to, but not a replacement for, classical testing methods. He also described the stage of development and use for each of the molecular technologies:

- PCR is in place, but remains challenging because of lack of standardization.
- Next-generation sequencing is still limited. Although not yet FDA approved, some nanopore sequencing is being used in the field.
- CRISPR-Cas is very promising but remains in the research phase.

[1] See Medical Dictionary, https://medical-dictionary.thefreedictionary.com/serology.

- MALDI is generally too expensive for low-resource settings.
- Multiplex PCR is available but is also expensive.
- Host response-based assays are likely to evolve rapidly in the future.
- Metagenomic sequencing is promising but also expensive and not yet widely available.

Dr. Chiu offered the following takeaway messages:

1. A combination of traditional methods (e.g., immunofluorescent strips, real-time PCR) and state-of-the-art approaches (e.g., nanopore sequencing, CRISPR-Cas assays, multiplexed PCR) will likely be needed moving forward.
2. Cost and other practical considerations favor true point-of-care molecular diagnostics (e.g., lateral flow immunoassays, CRISPR-Cas).
3. It will be important to decide whether the focus should be on diagnostic testing or surveillance. Who (*in loco*, in country, international) should be doing what? Emerging infectious diseases do not respect borders.
4. Sequencing has made the greatest impact in genomic surveillance, but not yet in molecular diagnostics.
5. MALDI, multiplexed PCR platforms (e.g., BioFire, Luminex), and even single-plex PCR instruments (e.g., Cepheid GeneXPert) remain too expensive for use in diagnosis, but may be acceptable for targeted surveillance, such as during outbreaks.
6. Inexpensive, field-ready multiplexed diagnostics are urgently needed but do not exist.
7. Direct detection approaches likely will not replace serology (e.g., lateral flow assays) anytime soon.
8. Complex data from genomic sequencing and other methods will require cloud computing resources to disseminate results quickly, which is critical in public health scenarios.

Dr. Chiu ended his presentation by stating that the effectiveness and accuracy of many of the molecular technologies must be demonstrated before they become widely usable in low-resource settings. Although some testing is occurring in low-resource settings, much work remains to be done.

During the discussion that followed, one participant said that his group is working to develop non-probe PCR techniques, trying to use multiplex immunoassays for serology, and providing Sanger sequencing using remote analysis, where needed, as a backup to PCR field applications. He noted that the real questions are how to deliver the new tools to the field and train people in all these new skills, especially data handling, storage, and security. Some of the tools now available require no maintenance and have disposable cartridges. One approach for data is to use cloud-based bioinformatics to analyze data and return results so that no local bioinformatics talent is needed for this purpose, provided internet connectivity exists. However, another participant stated that communications are a real problem in the field because bandwidth is insufficient. Therefore, cloud approaches may not work in an outbreak situation.

A participant asked whether a case can be made for aiming for reagent self-sufficiency. Dr. Chiu replied that there is because the reagent market is not very competitive and competition would likely drive down costs. The same participant said that a cost-benefit analysis, which does not exist but is needed, could help donors to decide what type of support they should provide.

With Dr. Chiu's summary of the state of the art and his conclusions about technology readiness, the participants were ready to examine the practicalities of field deployment and use. Jonathan Towner, PhD, of the U.S. Centers for Disease Control and Prevention (CDC) gave a presentation on his field experiences during several hemorrhagic viral outbreaks, including the West African Ebola outbreak in 2014-2015. Dr. Towner's experiences illustrate the application of available techniques to real-world responses to major disease outbreaks. His first field experience was in Uganda in 2000-2001, where CDC used both ELISA-based and PCR-based testing. The work was performed in a hospital lab, and more than 1,000 samples were processed over a 3-month period, with more than 280 testing positive for the virus. In 2005, he participated in the response to an outbreak of Marburg virus in Angola. ELISA and PCR were again used, and this time the work was performed in an existing lab that was established for HIV diagnostics. This time 180 of 505 samples from blood or serum, breast milk, or swabs were found positive for the virus. From 2010 to 2016, Dr. Towner participated in a program of enhanced viral hemorrhagic fever (VHF) surveillance and diagnostics in Uganda to provide training for the local medical and other staff. This program placed CDC personnel in-country and helped to achieve a greatly reduced number of later VHF cases as well as reducing the time to diagnoses.

Then in 2014, the huge Ebola outbreak in West Africa occurred, seriously affecting Guinea, Sierra Leone, and Liberia. This event attracted, in Dr. Towner's words, a "United Nations" of field response with

Germany, France, Italy, Belgium, the Netherlands, England, Canada, the United States, Nigeria, South Africa, China, and Russia sending people, equipment, and other aid in an impressive response and collaboration effort. The U.S. agencies included CDC, the National Institutes of Health, and the Department of Defense. The aim was to provide rapid diagnostics in the field. In all, 27 field laboratories were set up across West Africa.

The scale of the response to the West African Ebola outbreak generated its own challenges. For example, many different real-time PCR assays were used in the large network of laboratories, creating a need for quality panels and attempts to standardize assays. For the panels distributed by CDC in Sierra Leone and Guinea, 2 of 6 laboratories were producing incorrect results at a rate of 10 percent, which required implementation of improvements. The need for a two-target Ebola assay, plus cell RNA PCR controls, emerged to reduce the rate of false positives or negatives when only one target was used. The lack of a cell RNA control also increased the risk of false-negative results.

There were also database, documentation, and reporting challenges. It was difficult to complete sample submission forms, so a considerable number of samples had little or no documentation. There was an absence of unique identifiers for samples and cases and difficulties with linking lab, clinical, and epidemiological data. Information on date of onset was often missing, as was knowledge of whether swabs were from corpses (appropriate) or live patients (inappropriate). Finally, there were problems with turnaround time for results, insufficient numbers of trained phlebotomists, and transport of samples.

Dr. Towner concluded that much was achieved, despite these problems. Peak testing occurred during the October-December 2014 period, with the highest number of samples tested at 180 per day in July 2015. On average, 71 percent of samples were tested on the day they arrived at the lab, and 99.9 percent were tested either the same or the next day. Samples were received for about 14 months from 12 of 14 districts and were mainly whole blood and cadaver oral swabs. Overall, more than 27,000 samples were tested. Dr. Towner's lab remained operational for 406 days, with no days off or disruptions in testing. A pilot study testing for viral persistence in male survivors began on May 23 and resulted in the testing of more than 500 semen samples. The Sierra Leone vaccine trial, or STRIVE, began on May 24, and 51 samples from 30 participants were tested. Twenty-eight teams of personnel from 17 different branches throughout CDC were trained in Atlanta on the Bo lab protocols and procedures and then deployed to help keep the lab operational.

After the crisis phase of the epidemic was over, Dr. Aiah Lebbie of Njala University in Sierra Leone was selected as the recipient of a 3-year CDC cooperative agreement to conduct ecological VHF surveillance on the region's bats. The University's laboratory facilities underwent major lab renovations from March through August 2017. There are now stable and properly maintained electricity, freezers, and other working equipment. Purchasing is operational, although delivery of perishables remains an issue. This arrangement with the University will provide educational as well as public health benefits for the region and exemplifies what partnering can accomplish. Approximately 5,000 bat specimens have been collected, all from forested areas. There are two field stations, one on Tiwai Island in Sierra Leone, which is starting renovations, and another in Gola Rainforest National Park in Liberia.

Following Dr. Towner's presentation, one participant noted that the new molecular technologies reduce risk, so minimal containment levels are needed if pathogens such as Ebola and Marburg are indigenous to a particular locality. In such a case, BSL-3 and -4 laboratories are probably not needed, but some recipients seek high-containment facilities for prestige rather than real needs. Another participant restated the need to distinguish between labs conducting surveillance and those conducting diagnostics, and noted that reference labs are necessary for identification of strains and for research. Perhaps low containment is adequate for the field while higher containment is required for reference labs.

Although Dr. Chiu said that the costs of the molecular technologies need to be reduced before they can be used in low-resource settings, one participant stated that acceptable cost may depend on the disease, the number of affected patients, the costs of care and treatment, and other factors. This translates into common diseases needing cheaper analytical capabilities. Some technology is already spreading. For example, 165 "Gene Expert" machines have already been deployed throughout the Democratic Republic of Congo, although, as pointed out by one participant, their throughput is low. In addition, these machines only detect the Ebola Zaire strain, which, if it mutates, might be undetected like other strains. Another participant noted that funders should account for already-deployed capabilities when making support decisions.

A participant highlighted the different needs for normal operating situations vs. responses to epidemics. His organization uses Luminex testing, but what is appropriate depends on whether the researcher is looking for one specific pathogen or more. Another participant stated that anything new that is built should be linked to existing facilities that are already part of a network so that reagents and other resources can be shared.

Dr. Chiu stated that direct detection of the organism of interest is the "gold standard." Clinicians are more conservative than researchers, so it might be 5 to 10 years before new test types are accepted as the basis for patient treatment. Because nasty incidents with drug and vaccine testing have already occurred in developing countries, there is a particular need to be conservative with molecular diagnostics as new tools for guiding medical treatment on these grounds as well.

A participant reiterated that the cost of reagents is an issue, especially because of the small budgets of low-resource countries. Another participant pointed out that the technologies themselves are very costly and donors should be advised to rely on proven technologies as a baseline. In light of cost factors, another participant said that it makes sense to use distributed sets of labs for basic testing and to reserve the expensive, high-throughput capabilities for central locations. Yet another participant noted that emergencies are special, but emergency response will improve if more laboratories are equipped to deal with normal business. Having a lab in place for normal operation also keeps personnel and supply chains trained and practiced. Donors should also inquire about quality management systems. In developing countries, reference labs and other resources may not be available to facilitate standardization, verification, and other necessary steps.

A participant said that the "Wild West" can result from a lack of regulatory oversight—even in the United States, but more so in some developing countries. Another participant noted, however, that France provides oversight in Francophone countries. Another participant stated that the lack of reagents and of equipment maintenance and repair is a donor's nightmare and that training for these latter functions is greatly needed. However, another participant noted that the education levels in some localities are so low that the concept of maintenance does not exist and therefore providing such training for local personnel is very difficult. In addition, transportation poses a barrier to building reference capabilities. Partnerships are needed to provide fuel, dry ice, and other laboratory staples. Multinational corporations, such as Coca Cola, and oil companies, for example, have provided some of these supplies. The African Union, the Economic Community of West African States, and the South African Development Community have developed ways to address some of these problems,

A participant suggested that biobanks be sited in secure locations, such as military bases. Another participant noted that biobanks secure and recognize the value of live samples, but may not have plans for what to do with them, raising questions about whether they should have collections without clearly defined purposes.

Participants called for a broader vision to ensure preparation for the next outbreak. They also posed the question of how to interest donors in enhancing what already exists instead of building new facilities. Finally, they wondered how donors could help with "leave behind" facilities after an outbreak, as was done in Sierra Leone after the 2014 Ebola outbreak.

BOX 4.1

The State of Development of Molecular Diagnostics: A Summary

Dr. Charles Chiu described the state of development of molecular diagnositcs. He began by explaining that in "probe hybridization" or non-amplified nucleic acid probes, strands of DNA or RNA of less than 50 base pairs from a sample are probed for specific nucleic acid sequence "targets" that indicate the presence of particular pathogenic organisms. The DNA or RNA strands are labeled with enzymes, antigenic substrates, chemiluminescent molecular subunits, or radioisotopes. These bind with high specificity to complementary sequences of either DNA or RNA, which is referred to as "hybridization." The probes bind rapidly, under stringent conditions, and can detect even a single nucleotide change in a nucleic acid sequence. They can be used directly on patient specimens or on culture isolates. They are, however, 100-10,000 times less sensitive than amplification methods, and this level of sensitivity may not be sufficient for detecting organisms, such as Ebola virus, that have low copy number in tissue samples, such as blood. Probe hybridization is traditionally used when large numbers of organisms are present, although as noted the method is not very sensitive. It has been found particularly useful for confirming the presence of Mycobacteria species, systemic fungi, Campylobacter, Enterococci, Haemophilus influenzae, Group A and B Streptococci, Streptococcus pneumoniae, Neisseria gonorrhoeae, Staphylococcus aureus, and Listeria.

CRISPR-Cas based assays (CRISPR stands for Clustered Regularly Interspaced Short Palindromic Repeats and Cas for CRISPR-associated system) have received an enormous amount of recent publicity as a means of genome editing. These systems, derived from a natural bacterial defense mechanism against viruses that infect them, have made editing of the genome by adding, replacing, or removing DNA base pairs much more precise, efficient, flexible, and less expensive relative to previous strategies for changing gene sequences (National Academy of Sciences, National Academy of Medicine, 2017). They can be applied in non-human organisms, humans, and microorganisms. These advances have spurred an explosion of interest from around the globe in the potential ways in which genome editing can improve human health. Dr. Chiu stated that CRISPR-Cas-based molecular diagnostics could be a "game-changer" for the diagnosis of infectious diseases, surveillance of emerging pathogens, viral genotyping, detection of antibiotic resistance factors, cancer screening, and other applications.

CRISPR-Cas–based assays have advantages and disadvantages according to Dr. Chiu. Among the advantages are that these assays are highly sensitive, have turnaround times of less than 2 hours, and have great design flexibility, requiring less than 1 week from design to implementation. They are highly portable, do not require electricity or expensive instrumentation, and can provide colorimetric readouts that are relatively easy to interpret. The University of California at Berkeley recently demonstrated that human papillomavirus can be detected directly from genital tissue without extraction in 1 hour using CRISPR-Cas. It is also currently being used to identify Zika and dengue viruses in clinical samples. The disadvantages, however, include several factors related to the fact that CRISPR-Cas remains in the phase. It has unclear multiplexing capability, it requires a target amplification step, which could introduce contamination issues, and its performance in most actual clinical uses is not yet proven. There are also important regulatory issues as well as concerns about test availability because many intended uses may be under patent.

Dr. Chiu described signal and target amplification technologies. The most widely used signal amplification method for diagnostics is the Branched Chain Technology. This technology has long been used to determine viral loads for HIV and hepatitis C (Tsongalis, 2006). It can be used to detect proteins and nucleic acids, both DNA and RNA. The concentration of the probe or target does not change, but sensitivity increases with increased concentration of labeled molecules attached to the target nucleic acid. Its advantage over target amplification methods, such as PCR, is that the detection "signal" is directly proportional to the amount of the target in a specimen, allowing for easier quantification. There is also a decreased risk of contamination, and inhibitors are not a problem.

Target amplification methods, such as PCR and Transcription-Mediated Amplification (TMA), use enzyme-mediated processes to make copies of a target nucleic acid. The result is that the analyst gets millions or billions of the target, which can lead to problems with contamination and false-positive results. The World Health Organization has issued standards for PCR (WHO, 2016).

PCR is a relatively simple technique for amplifying and detecting DNA and RNA sequences, such those associated with the genetic material of pathogens. Compared to traditional methods of DNA cloning and amplification, which can take days to complete, PCR requires only a few hours. Double-stranded DNA is first heat denatured to separate the strands. Primers then align to the separated DNA strands (annealing), and DNA polymerase then extends the primers. The result is two copies of the original DNA strand. The denaturation, annealing, and elongation process constitutes one cycle of amplification, which is repeated 20-40 times. The amplified product can then be analyzed.

The method is widely used to amplify DNA for experimental use, genetic testing, and detection of pathogenic material. PCR is highly sensitive and requires only small sample volumes. Dr. Chiu highlighted a 2010 paper (Boehme et al., 2010) that showed that a PCR-based technique could rapidly detect the presence of TB, including antibiotic-resistant forms, from sputum samples. Boehme noted that global control of TB is hampered by slow and insensitive diagnostic methods, particularly for the detection of drug-resistant TB. Early detection reduces the death rate and interrupts transmission, but the complexity and infrastructural needs of sensitive methods limits their accessibility and effect in low-resource settings. The rapid detection test developed by Boehme et al. was quickly deployed in Africa and elsewhere, although it is expensive.

There have also been numerous instances of inter-lab discrepancies in the results of PCR, indicating a performance variation among labs, although PCR assays are validated, accredited, and routinely used in some labs. In addition, "genomic drift" impacts these assays and PCR performance deteriorates over time as viruses mutate. Multiplex PCR, which is a desirable but not yet fully realized goal, would allow analyses of multiple targets in the same sample. Dr. Chiu cited a paper by Mahoney et al. (2007) on the development of a multiplex PCR panel test for the detection of 20 human respiratory viruses.

Dr. Chiu's particular area of expertise is DNA sequencing. Many devices and methods for DNA sequencing currently exist. For example, for bacteria, 16S RNA forms part of the bacterial ribosome. These RNA fragments contain regions of highly conserved and "hypervariable" sequence that can be thought of as molecular "fingerprints" that can be used to identify bacterial genera and species. These conserved regions can be targeted to amplify a broad range of bacteria from clinical samples. However, the technique is not quantitative because of copy number variation. The major advantage of this approach is that there is no need for culture, so it can be used on a very broad range of organisms. In this approach, DNA is extracted from a clinical sample, such as tissue or body fluid. The 16S RNA genes are amplified, sequenced, and compared with a reference database, such as GenBank, to look for a match. This type of testing is just becoming available in clinical settings. A similar approach using the 18S, 28S, and ITS genes is also available for eukaryotic pathogens (fungi and parasites).

Another approach is matrix-assisted laser desorption ionization time of flight mass spectrometry (MALDI-TOF MS). This technology has emerged as a tool for identifying and diagnosing unknown microbes from intact cells or cell extracts using the proteins translated from microbial genomes. It is rapid, sensitive, and economical. It has been adopted by microbiologists for microbial identification and strain typing, epidemiological studies, and detection of biological warfare agents, water- and food-borne pathogens, antibiotic-resistance factors, and blood and urinary tract pathogens (Singhal et al., 2015).

Mass spectrometry (MS) involves ionizing chemical compounds into charged molecules and measuring their mass to charge (m/z) ratio. MS has been used since the early 1900s in the chemical sciences, but the development of electron spray ionization (ESI) and MALDI in the 1980s increased its applicability to large biological molecules, such as proteins.

Both ESI and MALDI are based on "soft ionization" methods that do not produce significant loss of sample integrity. In ionization by either, proteins are converted into ions by addition or loss of one or more protons. MALDI-TOF MS has certain advantages over ESI-MS because it produces singly charged ions, making data interpretation easy. In addition, prior separation by chromatography is required for ESI-MS, but not for MALDI-TOF MS, which is completely automated. The high throughput and speed associated with MALDI-TOF MS have made it superior for large-scale proteomics work.

The sample for analysis by MALDI-TOF MS is prepared by mixing or coating it with a solution of an energy-absorbent organic "matrix" compound. Both the matrix and the sample trapped in it are crystallized by drying, and the sample in the matrix is ionized with a laser beam. This generates protonated ions from the sample, which are then accelerated so that they separate from each other on the basis of their mass-to-charge (m/z) ratio. The ratio is measured by determining the time required for the ion to travel the length of the TOF tube (hence "time of flight"). Based on the TOF information, a characteristic spectrum called a peptide mass fingerprint (PMF) is generated for analytes in the sample. Microbes are identified by comparing the MS spectrum of an unknown microbial isolate to the spectra of known isolates in a reference database. Obviously, the limitation of the technology is that exact identification of new isolates is possible only if the reference database contains PMFs of the type strains of specific genera, species, and subspecies in the sample.

The topic of unknowns is very important in infectious disease. Dr. Chiu pointed out that of the top three common diseases—pneumonia, meningitis/encephalitis, and fever/sepsis—15-62 percent, 40-60 percent, and 20 percent, respectively, are caused by unknown organisms and are missed in hospital diagnostic labs even in developed countries. This makes the idea of being able to sequence everything in patient samples attractive, which can now be done using metagenomic next-generation sequencing. Metagenomics overcomes the twin problems of being unable to culture most microorganisms and dealing with the vast genomic diversity of microbes. These are the biggest roadblocks to advancement in clinical and environmental microbiology (National Research Council, 2007). Metagenomics seeks to understand biology at the aggregate level, transcending individual organisms and focusing on the genes in an entire microbial community, whether this is from a soil sample or the human body. It also requires the development of advanced computational methods that maximize understanding of the genetic composition and activities of communities so complex that they can be sampled, but never completely characterized (National Research Council, 2007).

Although still a relatively new science, metagenomics has produced much new knowledge about the unculturable microbial world using radically new ways of doing microbiology. All metagenomics work starts with the same first step: DNA is extracted directly from all the microbes present in a particular sample. This mixed DNA sample can then be analyzed directly or cloned into a form maintainable in culturable laboratory bacteria. This enables researchers to create a library of genomes of all the microbes found in that sample. This library can, in turn, be studied either by analyzing the nucleotide sequences of the cloned DNA or by determining what proteins the cloned genes make when they are expressed. The technique lends itself to sorting out complex disease situations that can have many different causes. For example, tropical febrile illnesses can be caused by numerous species of bacteria, viruses, or parasites while presenting similar symptoms, and the use of metagenomics can help determine which specific organism is causing the fever in affected patients.

A metagenomic library is analogous to thousands of jigsaw puzzles jumbled into a single box, and putting the puzzles together again is one of the great challenges of metagenomics. The approach is now possible because of the availability of inexpensive, high-throughput DNA sequencing and the advanced computing capabilities needed to make sense of the millions of random sequences contained in genome libraries. The latter requires a robust bioinformatics pipeline, such as the Sequence-based Ultra-Rapid Pathogen Identification (SURPI) pipeline referenced by Dr. Chiu.

As an example of a metagenomics technology that might be suitable for use in low-resource settings, Dr. Chiu discussed nanopore sequencing, which allows for real-time metagenomic pathogen detection in patients with fever/sepsis. The advantages of nanopore sequencing include its ability to perform real-time sequence analysis, long reads, and direct sequencing of DNA, RNA, and protein from samples. It is portable using a pocket-sized device, does not require internet connectivity, and offers potentially fast turnaround times, which are key for infectious disease sequencing. Its disadvantages are that its use is costly ($500 per flow cell), it still has error rates of 8-12 percent, and the quality of its flow cells can be variable. Still, nanopore sequencing has been used successfully for real-time Ebola surveillance in West Africa (Quick et al., 2016). Quick, et al. show that they generated results in less than 24 hours after receiving an Ebola positive sample, with the sequencing process taking as little as 15-60 minutes. This illustrates that real-time genomic surveillance is possible in low-resource settings and can be established rapidly to monitor outbreaks. Dr. Chiu's group at the University of California, San Francisco has also published a recent paper (Thézé et al., 2018) that describes the use of metagenomics to reconstruct the introduction and spread of Zika virus in Mexico and Central America.

In another paper cited by Dr. Chiu, Gardy and Loman (2017) state that:

> "The recent Ebola and Zika epidemics demonstrate the need for the continuous surveillance, rapid diagnosis and real-time tracking of emerging infectious diseases. Fast, affordable sequencing of pathogen genomes—now a staple of the public health microbiology laboratory in well-resourced settings—can affect each of these areas. Coupling genomic diagnostics and epidemiology to innovative digital disease detection platforms raises the possibility of an open, global, digital pathogen surveillance system. When informed by a One Health approach, in which human, animal and environmental health are considered together, such a genomics-based system has profound potential to improve public health in settings lacking robust laboratory capacity."

There are, nevertheless, challenges to realizing this potential. Gardy and Loman describe some of these challenges for Zika virus, which exhibits low viral titers, a small genome (<11 kilobases), and transient viremia. Taken together, these factors complicate the detection of viral nucleic acid by a metagenomic approach. Gardy and Loman also report that obtaining a sufficient amount of viral nucleic acid for genome sequencing beyond simple diagnostics may also require PCR and an amplicon sequencing approach. Other challenges may include "access to reliable Internet connections, the ability to collect sample metadata, and translating genomic findings into real-time, actionable recommendations."

5

KEY FACTORS FOR BUILDING AND SUSTAINABLY OPERATING HIGH-CONTAINMENT LABS IN LOW-RESOURCE CONTEXTS: AN OVERVIEW

Dr. Nancy Connell, a member of the organizing committee, gave a presentation on key factors for building and sustainably operating high-containment labs in low-resource settings to highlight what determines success. She began with a list of desirable factors, all of which were mentioned during previous presentations or discussions:

- Appropriate infrastructural components and adequate budgets and supply chain for power, water, equipment, reagents, labor, and maintenance services;
- Management and administrative controls and culture;
- Mechanisms to counter safety and security threats;
- Regulatory framework, standards, and enforcement mechanisms;
- Effective regular inspections;
- Affiliation with biosafety and biosecurity organizations, curricula, and training to ensure professional competency; and
- Multidisciplinary design and execution.

Dr. Connell then explained that in 2015, the U.S. government submitted a document to the Meeting of Experts for the Biological Weapons Convention (BWC) titled "The United States of America High Containment Laboratory Policy." This document, which was included in the meeting book for participants, laid out five guiding principles:

1. Establish a demonstrated need for high-containment biocontainment facility in the country.
2. Establish that the recipient has demonstrated the commitment and ability to operate and maintain the facility upon completion.

3. Establish that the recipient country demonstrates commitment to nonproliferation.
4. Foreign policy considerations are addressed.
5. Factors related to biological risk management are considered (biosafety, security, training, local codes, and regulations).

Dr. Connell summarized the key points of the U.S. policy document as follows and elaborated on each of these elements.

Sustainability: The operation and maintenance of a high-containment laboratory is an extraordinarily expensive process. Moreover, although required, reliable, high-quality infrastructure (e.g., power, water, waste handling), replacement parts, and trained maintenance and repair personnel may not be readily available in some areas.

"Fit": "Fit" should be interpreted in terms of national priorities and needs. The report *Biosecurity Challenges of the Global Expansion of High-Containment Biological Laboratories* (National Research Council, 2012) noted that "when contributing to a new laboratory, donor groups and national governments do not . . . always ascertain how the new facility will complement other existing and planned infrastructure" (page 9).

Safety: High-containment laboratories may increase safety, but only when accompanied by ongoing training, adherence to appropriate protocols, procedures, regulations, and guidelines, and oversight.

Nonproliferation (biosecurity): Although there is a legitimate need for biocontainment facilities worldwide, the inherent dual-use potential of these facilities and related equipment, as well as the pathogens they contain and the skills developed through hands-on work, merit scrutiny in a world where terrorism and the proliferation of weapons-relevant materials, technologies, and expertise pose genuine threats.

The purpose of this brief discussion of the U.S. policy document was to prepare the plenary group for a breakout session to discuss the basis for developing a candidate set of "norms" for funding high-containment laboratories. Dr. Connell shared the following list of "factors to consider," which combine the key factors and policy considerations, to focus the breakout discussions:

- Assessment of site-specific challenges and needs
- Obtaining commitments of support from in-country government officials

- Soundness and availability of necessary infrastructure: power, water, transportation network, waste treatment, and communications
- Presence or absence of country- or region-specific regulatory framework guidelines and/or standards
- Availability of appropriately trained and credentialed local workforce
- Access to a national or regional biosafety organization
- Biosecurity and nonproliferation considerations.

SUMMARY OF BREAKOUT SESSIONS

The group divided into two breakout groups in separate rooms. Breakout Group 1 was chaired by Ann Arvin, with Fran Sharples as the rapporteur. Breakout Group 2 was chaired by David Franz, with Nancy Connell as the rapporteur.

When the groups returned to the plenary room, Group 1 reported out first, with Dr. Sharples summarizing the group's discussion. Starting with the establishment of needs, participants believe that the selection process is very important, and funders could consider narrowing their focus to fewer but better projects to stretch resources, because not every project can be funded. However, to establish needs, must one follow a specific process of just answering questions? The crucial question is "What is the lab for and why is it needed?" It is also important to understand what is driving the request, that is, "which scientists want to perform research or other work in the laboratory? Who is committed to the project, and are in-country national government officials committed?" If the answer to the last question is "no," can funders help to motivate them to commit to the project? One participant said that a strong commitment from the *local* authorities, who may have some control over finances or infrastructure, is also essential. Finally, does the laboratory have a champion, that is, someone who can effectively gather support for the project?

One participant said that funders should distinguish between the scientific, political, and financial factors underlying a request for a laboratory. Another participant added that the "branding" of laboratories with BSL-2 or -3 labels can cause problems, stimulating requests that over-reach needs and capabilities. A third participant commented that funders should shift from thinking "donation" to thinking "investment." Funders should view a project in terms of a cooperative agreement from the beginning and should engage potential multi-sectoral partners as early as possible—for example, local universities and companies. Funders could also think in terms of how a new or enhanced laboratory could serve as a "nucleation point" in a national or regional network.

Having the necessary infrastructure is a guiding principle. Infrastructure includes not only water supply and quality, power, waste treatment (including incinerators), and transportation resources, but also telecommunications and internet connectivity. For a new laboratory, electricity is perhaps the most important resource. The frequency of past outages and voltage fluctuations should be checked before any design plans are made, and backup generators and regulators should be provided if necessary, a participant said. Water quality is also very important, and a sound transportation network is critical for moving samples and supplies. This last need may be the hardest to deal with because there are not many good solutions if the existing network is inadequate. Another person familiar with a number of labs did indicate, however, that there are successful solutions that could be examined as potential models. Waste treatment, including incinerators, is also essential.

Once built, a laboratory's operational needs are diverse and include funding, equipment, reagents, and human resources. A partnership among many funding organizations might be required to sustain operations. Realistically, recipient countries are rarely positioned to assume responsibility after the 3- to 5-year planning, construction, and startup period, and partnerships can help here. A shift among funders to plan for the longer term (e.g., 10-15 years) might facilitate longer and broader commitments. Japan created some successful partnerships with Zambia, for example, that have been sustained over decades because of Japan's continued commitment to financial and technical assistance. A participant suggested that comprehensive transition plans, rather than simple handoffs, are needed. Continued involvement by funders or other supporters could also help to protect against proliferation threats. Co-ownership of labs could be more widely applied—if not interpreted as infringing on the sovereign rights of recipient countries.

The group recommended that funders have a "risk list" to review when making funding decisions. This list includes laboratory needs that could result in bad consequences if they were unreliable, including operations (e.g., power, water, security); training (i.e., is there a minimum standard for clinical labs or universal precautions?); retention and a pipeline for new hires to replace trained employees who leave;[1] financial sustainability; internal monitoring and assessment; buy-in and ownership from the

[1] Training and experience often result in more opportunities for the staff to work elsewhere.

national government (e.g., is there a clear commitment, such as a line item in the national budget?); and potential for collaboration (e.g., are the negotiators and lab officials able to work well with the local community and who is involved?). Participants said that strategies to address all of these risk items must be flexible and adapted to circumstances. Funders might want to "map" these risks.

The relationship between the funder and recipient should be based on trust. Many recipient countries may be sensitive to words such as "audit," which is used by some funders to mean project evaluation and progress checks. Financial audits are expected, but audits of the relationship between the funder and the recipient may construed as offensive. Some funders call these types of evaluations "supervisory assessments" and use a checklist to determine compliance with contracts. Development banks and other funding organizations require both internal and external audits because they give money to recipient governments to spend, and there have been too many past examples of funds being diverted. One participant mentioned that cooperative agreements from the U.S. Centers for Disease Control and Prevention (CDC) require both technical and financial audits, which are generally welcomed because they protect the lab director. CDC also works with the recipient's staff, teaching and improving biosafety and standard operating procedures. One participant mentioned, however, that donors are not homogeneous as to how they operate. Providing funds for the recipient country to contract, build, and operate a laboratory is very different from providing both funds and all the other necessary capabilities to a project.

One participant stated that many funders know what to consider in the other domains, but they need guidance on biosecurity and nonproliferation. A meeting on governance of dual-use research in the life sciences held by the U.S. National Academies of Sciences, Engineering, and Medicine (the National Academies) in June 2018 in Croatia highlighted that this area of governance is a very low priority, if at all, for many countries. Another participant agreed the BWC receives very low priority, if any, in many recipient countries, who do not deem bioterrorism to be a risk. In response to the need to raise awareness, the attendees of the recent (June 2018) governance workshop identified training on biosecurity issues as one answer. Another participant expressed more concern about insider threats, which are not addressed in many donation agreements. Dealing with potential insider threats requires good management, leadership, knowledge of the laboratory personnel, and development of a culture of responsibility among laboratory personnel. High-containment laboratories built by Japan in other countries are provided "maximum measures" for physical security, and personnel are trained in Japan on both biosafety and biosecurity while they are given other training for other issues for other procedures and operations. Personnel are asked to develop biosecurity guidelines and then implement them at the facility.

A participant noted that physical security protects a lab from outside threats, while "biosecurity" focuses more on internal threats. In addition, "biosecurity" means different things to different people. Another participant mentioned that funders could use current bio-risk management frameworks as guidelines for awareness training. However, a participant noted that context also matters. A regional laboratory working with Ebola virus must worry about biosecurity. In comparison, a field laboratory dealing with rat urine contaminated by Lassa fever in an area where 10 percent of the rats carry the disease will not worry about biosecurity, although biosafety is a concern. Security and safety are not always linked and should not always be linked, a participant said. The presence of hazardous pathogens in a particular location influences not only how biosecurity is defined for that locale, but also how the need for a laboratory is defined.

According to Dr. Connell, the two breakout groups had similar discussions and observations. Group 2 also discussed the importance of establishing the need for a lab—what does a recipient need and why? A laboratory's purpose may be for diagnostics, for surveillance, as a repository for preserving or banking strains, or for research. Are the recipient's diagnostic needs related to day-to-day health and medicine and/or outbreak response? Do the scientists plan to use traditional microbiology or newer molecular techniques? Do they understand the potential costs and complexity? What about the proposed requirements for training? Are local candidates available to fill the positions? The answers to these questions will dictate the type of laboratory needed. All of these things influence what kind of lab capabilities might be required to meet recipient needs.

Group 2 produced a list of various "models" of assistance for biological laboratories for low-resource countries based on the experience of the participants:[2]

U.S. Department of Defense model: Construction of large and fully equipped high-containment laboratories and peripheral laboratory

[2] These are simplified sketches created to communicate overall approaches and are not meant to characterize everything that a particular nation or organization does.

networks (e.g., Tbilisi and Almaty); may encompass work on both human and animal pathogens.

Canadian Custom Package model: New or repurposed modular units; two have been created in Nigeria and Sierra Leone; they provide safety with reduced complexity and are peer designed using core lab specifications plus or minus specific functions.

Japanese model: Upgrade of existing high-containment laboratories (*not* new construction) as part of creating a network; they also provide extensive training and education over the decades-long duration of the partnership.

French model (Pasteur): Maintain long-term (Pasteur Institutes date back 130 years) established labs; provide their own skilled personnel as well as relatively long education/training of local personnel.

Zambia model: Six agent-specific BSL-3s, each devoted to a different specific pathogen; these "Container Labs" are managed by the Zambia AIDS Related Tuberculosis Project, which is affiliated with the University of Zambia's School of Medicine and the London School of Hygiene and Tropical Medicine.

World Bank model: The following areas are important for consideration by donors, with the ones in bold of highest importance.

- Political and governance (including corruption)
- Macro-economics
- **Sector strategies and policy**
- **Technical design of project**
- **Institutional capacity and sustainability**
- Fiduciary risk
- Environmental and social risk
- Stakeholder commitment (are they on board?)

There was considerable discussion of obtaining "buy in" from both local and national government officials. Two participants said that involvement of the recipient country's minister of health may no longer be sufficient. Decision making now rests with the vice president or the Cabinet in some places.

Participants discussed the pros and cons of One Health. Laboratories that work on human and animal health are "double track," providing resources for two purposes with a single investment. This can be important for dealing with zoonotic threats, but stored agents should be kept separate. However, a participant stated that human health and animal health compete for resources when scarce.

Breakout Group 2 also produced a list of "Donor Operational Risk Analysis Factors:"

- Human resources: staff, training and education
- Ownership/buy in
- Collaboration and how well a country works with others
- Utilities
- Operations
- Facilities
- Finance and financial history
- Monitoring
- Regulatory framework
- Security

The importance of training and education received a lot of attention from both groups. One participant suggested that a donor agreement should require a training program that continuously serves the needs of a laboratory and its host country. Donors, local or foreign educational institutions, or national or regional biosafety professional organizations can provide training. Training should also encompass mechanisms for employee retention, motivation, and mapping of career paths. Donors need to consider alternative ways to deal with these needs.

Breakout Group 2's takeaway messages were as follows:

- The conversation about the different models should have started 20 years ago.
- The donor risks that were identified were not surprising.
- Forcing the use of culture-free techniques is infeasible.

6

POTENTIAL NORMS FOR FUNDERS OF BIOLOGICAL LABORATORIES IN LOW-RESOURCE COUNTRIES

In the final plenary session, the group discussed suggestions from many participants for "norms" for donors. As noted in the introduction to these proceedings, the views contained herein are those of individual workshop participants and do not represent the views of all workshop participants, the planning committee, or the National Academies of Sciences, Engineering, and Medicine.

ESTABLISHING NEEDS

No participant suggested that high-containment laboratories should not be constructed in low-resource settings. High-containment labs may be needed for many reasons, including performing diagnostics, supporting clinical care, supporting epidemiology, identifying cases and chains of transmission, supporting and carrying out ecological investigations, and conducting applied and basic research.

However, participants emphasized the need to match the facility type with the country's needs as well as its competency to operate and maintain the facility. Donors, a participant said, should be wary of ambitions that exceed capabilities. Funders must ask as many of the "right" questions as possible to determine the true requirements. Recipient countries must be able to justify the need for a laboratory and to explain the work that will be performed there. Conversations to establish answers to these questions may take a long time and may never reach resolution.

Several participants also strongly believe that if the "right" facility for a country is less than a BSL-3, then that is all that a funder should provide. Matching the appropriate type, design, and level of facility to the needs of the recipient is the focus of a joint effort by the government of Canada, the

World Organisation for Animal Health, and Chatham House. Some participants noted that countries where very hazardous pathogens, such as Ebola virus, are endemic may not need high-containment facilities to do diagnostic and clinical work because the organisms are already present in the surrounding environment.

CONSTRUCTING NEW LABS VERSUS REPURPOSING EXISTING ONES

Phased approaches that increase lab capability as competency increases might be a beneficial approach when a recipient asks for a high-containment laboratory that exceeds its needs and capabilities at that time. This would avoid the issue of donors committing to provide a high-containment laboratory before the recipient is ready to handle it. However, it is technically difficult to renovate a laboratory designed for BSL-2 into a BSL-3 laboratory. If a donor wishes to encourage tiering or repurposing, then the original lower-level laboratory should be designed with the intent to add on to it to facilitate repurposing success.

LABORATORY NETWORKS

Donors might also want to consider how a new laboratory might expand or complement existing national and/or regional capabilities, some participants noted. What capabilities are already in place? Have existing laboratories been operated as intended and sustained over time? Are there ways in which a new or enhanced facility could complement the capabilities of existing labs or leverage resources by sharing?

RISK ASSESSMENT

Risks for donors: Serious issues arise when funding is insufficient to support ongoing operations, the necessary infrastructure is inadequate, no regulatory framework exists, government officials fail to provide support, the availability of an appropriately trained and skilled workforce is questionable, and awareness of biosecurity and nonproliferation concerns is insufficient. Due diligence on the part of donors requires that an accurate picture of the situation in each of these areas be obtained before committing funding or approving a particular facility design.

Risks for workers and the local community: Personnel must learn, understand, and follow all laboratory safety protocols to ensure that infectious live pathogens do not cause illness for themselves, their colleagues, and members of the local community outside the laboratory walls. Donors cannot simply fund the construction of a lab and then walk away, assuming that sustainment of safety, security, and effectiveness will take care of themselves; there is ample evidence that this is not the case. The degree of risk and the containment level needed to ameliorate that risk depend on the types of pathogens being handled, their mode of transmission, whether effective countermeasures exist, and the factors discussed by Drs. Ksiazek and Kojima and summarized in Chapter 2. Although most donors may not be responsible for day-to-day lab operations and maintenance, they must perform due diligence and create contracts or other funding agreements that recognize these crucial concerns.

ALTERNATIVE TECHNOLOGIES

The new molecular diagnostic approaches now under development hold the promise of being able to substitute faster and safer surveillance and diagnostic testing for the more classical approaches that require handling of viable infectious organisms (see Chapter 4). Unfortunately, participants observed, very few of these technologies are ready for use even in sophisticated laboratories in wealthy countries, much less in low-resource settings. Their specificity is a key strength in positively identifying a particular strain. However, the specificity is also a weakness in that their utility is compromised when strains of a pathogen are different or mutating. In addition, few of the new technologies have been standardized or validated, especially under field conditions, and their effectiveness and accuracy must be demonstrated before they can be widely used in medical diagnostics. In addition, their costs must be reduced if they are to be used in low-resource settings. Finally, none has received regulatory approval. Dr. Chiu stated that it might take 5 to 10 years before the new test types are accepted as a basis for patient treatment. It is, therefore, unrealistic at this time to dismiss the need for biological containment facilities based on a belief that alternative technologies will obviate the need for handing of live infectious organisms, at least in some circumstances.

HUMAN VERSUS ANIMAL HEALTH

Although participants recognized that One Health is a good approach to protect both human and animal health, there are few examples in developing countries of human and animal pathogens being handled at the same facilities. In the United States, however, facilities such as the Galveston, Texas, BSL-4 lab may host work on both, in different spaces. When this is the case, the human and animal organisms are generally kept well separated. This precaution is intended to reduce the chances of organisms, such as avian influenza, coming into contact with human flu strains and undergoing reassortment to become infectious to humans. The World Health Organization (WHO) prohibits handling both animal and human strains of influenza in the same laboratories. Aside from such specific exceptions, however, a facility suited for work on Level 3 human pathogens is generally also suitable for work on Level 3 animal pathogens, which could be useful for surge capacity.

Several participants noted that developing countries are usually more interested in public health and that the agricultural sector does not receive nearly the same level of funding and attention. Typically, the lines of authority for human and animal health are separate, which is a complicating factor.

REGULATORY FRAMEWORK

The International Health Regulations (IHR) are legally binding on 196 Member States worldwide. They impact governmental functions and responsibilities across many ministries, sectors, and governmental levels in many countries. However, WHO does not specify how the legal and regulatory requirements imposed by the IHR are to be implemented. It is up to each State Party to do so in the context of its own legislation, governmental structures, and policies (WHO, 2009). The effective implementation of IHR obligations, however, requires that an adequate legal framework be in place. In some Member States, the relevant authorities adopt implementing legislation. Although new or revised legislation, regulations, or other instruments may not be explicitly required under the State Party's legal system, a country may still consider adopting them to facilitate performance of IHR activities in a more efficient and effective way.

However, approximately only one-half of the 196 Member States met the 2016 deadline for IHR implementation. Consequently, many low-resource countries still lack formal legal and regulatory frameworks that apply to the operation and management of the biocontainment laboratories they need to protect public health. In these cases, prospective donors should seek alternatives, such as requiring the recipient country to use existing guidelines or best practices, such as the WHO Laboratory Biosafety Manual or CDC's Biosafety in Microbiological and Medical

Laboratories manual. The recipient country's willingness to commit to this requirement should factor into the donor's decision-making process.

The various biosafety manuals also require regular inspections of equipment and other aspects of lab operation and management to ensure that safety and security measures are given appropriate attention. One participant noted, however, that the newer types of equipment are modular and may require maintenance by the manufacturer, rather than the lab. A surveillance system to detect infections among lab workers is also required. Finally, some participants noted that financial audits are needed to prevent corruption and diversion of funds to unacceptable purposes. Supervisory assessments using checklists can be used to "audit" operations. Verification and enforcement of whatever rules are in place will be necessary and might require contracting to a third party.

INFRASTRUCTURE

As noted by many participants, reliable power, water, transportation, resupply, waste treatment, telecommunications, and internet connectivity are all needed, if not absolutely required, to ensure that biological laboratories operate for the purposes for which they were constructed. Donors should ensure that these resources are available, and funding agreements should specify how these resources will be provided and by whom. Furthermore, donors should gather data on the reliability of these resources, perhaps by spending time on site and devising corrective measures if needed. How well the issues of availability and reliability are addressed elsewhere in the country could signal what to expect at a new laboratory site.

Donors should also determine in-country support for infrastructure. That is, Will the regional or national government contribute to the success of the project by contributing new or improved infrastructure components? Will the government accept some financial responsibility for maintaining new infrastructure paid for by the donor?

BIOSAFETY AND BIOSECURITY AND NONPROLIFERATION

Biosecurity has received little attention in low-resource countries. It is, nevertheless, a crucial and element in the operation and maintenance of biological laboratories as is biosafety. The legal framework for biosecurity includes the requirements imposed by the Biological Weapons Convention, the Australia Group, and U.N. Resolution 1540. The WHO, CDC, and other manuals provide guidance on implementing biosafety and biosecurity requirements. The June 2018 National Academies workshop in Zagreb, Croatia, on governance concluded that addressing biosecurity and the need to prevent access to hazardous pathogens by people with malicious intent requires awareness and procedural training for laboratory management and staff. A participant said that donors must ensure that the recipient recognized these topics are important and has a plan to implement the required measures. Some participants suggested the use of bio-risk management plans, which integrate safety and security. Donors should look for evidence that lab management and governmental authorities are aware of all safety and security needs and, if not, seek commitments from recipients to create awareness in all persons who need it.

WORKFORCE AND TRAINING

The need for trained workers who are competent to handle pathogenic organisms was mentioned numerous times during the workshop. Key points made in the discussions included:

- Education and training are not the same thing. Low levels of education in the locality of a proposed new laboratory may make it difficult to ensure that local people can be adequately trained to competently carry out laboratory procedures.
- Local sources of training for lab workers may not be readily available depending on the location. Donors could contract with third parties to provide training, or, if a technical organization itself, could provide training by its own personnel. In some cases, personnel may receive training in facilities in the donor nation, which enables them to experience the entire operation, including its culture. Professional societies and international biosafety organizations are other potential sources of training.
- The management of a lab workforce involves more than just training employees to carry out their responsibilities. Measures to assist employees with acquiring leadership skills and with career planning, as well as providing promotion opportunities, can reduce attrition and prevent experienced workers from accepting higher level jobs at other facilities.
- A pipeline to counteract attrition among workers should be developed. Programs such as internships could create a pool of potential employees who are prepared to enter a lab's workforce. Collaborative relationships with local colleges and universities could provide a

source of staff with relevant experience. Such relationships can also improve access to training by educational institutions.

- Collaboration with foreign universities is a potential means to provide training, including in more advanced scientific and technological subject matters. Some low-resource countries already have relationships with such foreign institutions (e.g., Zambia).

Dr. Arvin closed the workshop with thanks from the National Academies.

REFERENCES

Boehme, C.C., Nabeta, P., Pamela, N., Hillemann, D., Nicol, M.P., Shenai, S.D., Krapp, F., Allen, J., Tahirli, R., Blakemore, R., Rustomjee, R., Milovic, A., Jones, M., O'Brien, S.M., Persing, D.H., Ruesch-Gerdes, S., Gotuzzo, E., Rodrigues, C., Alland, D., and Perkins, M.D. 2010. "Rapid Molecular Detection of Tuberculosis and Rifampin Resistance." *The New England Journal of Medicine*, 363:1005-1015. doi: 10.1056/NEJMoa0907847.

Clinical and Laboratory Standards Institute (CLSI). 2015. Molecular Diagnostic Methods for Infectious Diseases, 3rd Edition. Available at https://clsi.org/standards/products/molecular-methods/documents/mm03.

Flint, M., Goodman, C.H., Bearden, S., Blau, D.M., Amman, B.R., Basile, A.J., Belser, J.A., Bergeron, É., Bowen, M.D., Brault, A.C., Campbell, S., Chakrabarti, A.K., Dodd, K.A., Erickson, B.R., Freeman, M.M., Gibbons, A., Guerrero, L.W., Klena, J.D., Lash, R.R., Lo, M.K., McMullan, L.K., Momoh, G., Massally, J.L., Goba, A., Paddock, C.D., Priestley, R.A., Pyle, M., Rayfield, M., Russell, B.J., Salzer, J.S., Sanchez, A.J., Schuh, A.J., Sealy, T.K., Steinau, M., Stoddard, R.A., Taboy, C., Turnsek, M., Wang, D., Zemtsova, G.E., Zivcec, M., Spiropoulou, C.F., Ströher, U., Towner, J.S., Nichol, S.T., and Bird, B.H. 2015. "Ebola virus diagnostics: The U.S. Centers for Disease Control and Prevention Laboratory in Sierra Leone, August 2014 to March 2015." *Journal of Infectious Diseases* 212 (Suppl 2):S350-3588. doi: 10.1093/infdis/jiv361. Epub 2015 Jul 30.

Gardy, L., and N. Loman. 2018. "Towards a genomics-informed, real-time, global pathogen surveillance system." *Nature Reviews Genetics* 2018 (19):9-20. doi: 10.1038/nrg.2017.88. Epub 2017 Nov 13.

Kimman, T.G., Smit, E., and Klein, M.R. 2008. "Evidence-based biosafety: A review of the principles and effectiveness of microbiological containment measures." *Clinical Microbiology Reviews* 21(3):403-425.

Kojima, K., Makison Booth, C., Summermatter, K., Bennett, A., Heisz, M., Blacksell, S.D., and McKinney, M. "Risk-based reboot for global lab biosafety: New WHO guidance could expand access to lab facilities." 2018. *Science* 360:260-262. doi: 10.1126/science.aar2231.

Mahony, J., Chong, S., Merante, F., Yaghoubian, S., Sinha, T., Lisle, C., and Janeczko, R. 2007. "Development of a respiratory virus panel test for detection of twenty human respiratory viruses by use of multiplex PCR and a fluid microbead-based assay." *Journal of Clinical Microbiology* 45:2965-2970.

NRC (National Resource Council). 2012. *Biosecurity Challenges of the Global Expansion of High-Containment Biological Laboratories.* National Academies Press. Washington, D.C. Available at: https://www.nap.edu/catalog/13315.

Quick, J., Loman, N.J., Duraffour, S., Simpson, J.T., Severi, E., Cowley, L., Bore, J.A., Koundouno, R., Dudas, G., Mikhail, A., Ouédraogo, N., Afrough, B., Bah, A., Baum, J.H., Becker-Ziaja, B., Boettcher, J.P., Cabeza-Cabrerizo, M., Camino-Sanchez, A., Carter, L.L., Doerrbecker, J., Enkirch, T., Dorival, I.G.G., Hetzelt, N., Hinzmann J., Holm, T., Kafetzopoulou, L.E., Koropogui, M., Kosgey, A., Kuisma, E., Logue, C.H., Mazzarelli, A., Meisel, S., Mertens, M., Michel, J., Ngabo, D., Nitzsche, K., Pallash, E., Patrono, L.V., Portmann, J., Repits, J.G., Rickett, N.Y., Sachse, A., Singethan, K., Vitoriano, I., Yemanaberhan, R.L., Zekeng, E.G., Trina ,R., Bello, A., Sall, A.A., Faye, O., Faye, O, Magassouba, N., Williams, C.V., Amburgey, V., Winona, L., Davis, E., Gerlach, J., Washington, F., Monteil, V., Jourdain, M., Bererd, M., Camara, A., Somlare, H., Camara, A., Gerard, M., Bado, G., Baillet, B., Delaune, D., Nebie, K.Y., Diarra, A., Savane, Y., Pallawo, R.B., Gutierrez, G.J., Milhano, N., Roger, I., Williams, C.J., Yattara, F., Lewandowski, K., Taylor, J., Rachwal, P., Turner, D., Pollakis, G., Hiscox, J.A., Matthews, D.A., O'Shea, M.K., Johnston, A.M., Wilson, D., Hutley, E., Smit, E., Di Caro, A., Woelfel, R., Stoecker, K., Fleischmann, E., Gabriel, M., Weller, S.A., Koivogui, L., Diallo, B., Keita, S., Rambaut, A., Formenty, P., Gunther, S., and Carroll, M.W. 2016. "Real-time, portable genome sequencing for Ebola surveillance." 2016. *Nature* 530:228-232. doi: 10.1038/nature16996. Epub 2016 Feb 3.

Shoemaker, T., Balinandi, S., Tumusiime, A., Nyakarahuka, L, Lutwama, J., Mbidde, E., Kofman, A., Klena, J.D., Ströher, U., Rollin, P.E., and Nichol. S.T. 2018. "Impact of enhanced viral haemorrhagic fever surveillance on outbreak detection and response in Uganda." *The Lancet Infectious Diseases* 18: 373-375.

Singhal, N., Kumar, M., Kanaujia, P.K., Virdi, J.S. 2015. MALDI-TOF mass spectrometry: "An emerging technology for microbial identification and diagnosis." *Frontiers in Microbiology*. 6:791. doi: 10.3389/fmicb.2015.00791. eCollection 2015.

Thézé, J., Li, T., du Plessis, L., Bouquet, J., Kraemer, M.U.G., Somasekar, S., Yu, G., de Cesare, M., Balmaseda, A., Kuan, G., Harris, E., Wu, C.H., Ansari, M.A., Bowden, R., Faria, N.R., Yagi, S., Messenger, S., Brooks, T., Stone, M., Bloch, E.M., Busch, M., Muñoz-Medina, J.E., González-Bonilla, C.R., Wolinsky, S., López, S., Arias, C.F., Bonsall, D., Chiu, C.Y., and Pybus, O,G. 2018. "Genomic epidemiology reconstructs the introduction and spread of zika virus in Central America and Mexico." *Cell Host Microbe* 23(6):855-864.e7. doi: 0.1016/j.chom.2018.04.017. Epub 2018 May 24.

Tsongalis, G.J. 2006. "Branched FDNA technology in molecular diagnostics." *American Journal of Clinical Pathology*. 126:448-453.

World Health Organization. 2009. International Health Regulations (2005): Toolkit for implementation in national legislation. International Health Regulations Coordination, Health Security and Environment, World Health Organization. Geneva, Switzerland. Available at http://www.who.int/ihr/NFP_Toolkit.pdf?ua=1.

World Health Organization. 2016. Establishment of PCR Laboratory in Developing Countries, second edition. Printed in India. Available at http://www.searo.who.int/entity/bloodsafety/pcr_lab_2016.pdf?ua=1.

APPENDIX A

STATEMENT OF TASK

The Statement of Task for the workshop in Amsterdam was as follows:

An ad hoc planning committee will organize an international workshop on guiding principles and common norms for use by countries, organizations, and experts who donate or provide biological laboratories for public and animal health in low-resource contexts.

The workshop will engage both U.S. and major international donors who fund establishment of biological laboratory capacity in developing countries; representatives of major development banks; representatives of major foundations that fund laboratories; representatives of laboratory design, engineering, and construction firms; other national scientific academies; and experts who perform diagnostics, conduct research, and lead response to outbreaks of dangerous pathogens in low-resource contexts, to identify guiding principles and common norms that they believe should inform donors as they consider laboratory projects. By serving as a neutral convener, the National Academies of Sciences, Engineering, and Medicine will lay the foundation for development of norms for provision of laboratory capacity.

The workshop will include: (1) an overview of the 2011 National Academies workshop held in Istanbul on high-containment biological laboratories and developments since then, focused on sustainability challenges with high-containment laboratories (BSL-3s and BSL-4s) in low-resource contexts; (2) models of successful provision of lower-containment laboratory capacity that meet diagnostic and research needs, including a review of the current state of the science of culture-free diagnostic methods and recent experience in West Africa; and (3) nonproliferation, biosecurity, and biosafety considerations. The workshop will also identify topics to explore during future meetings.

A rapporteur will produce a brief summary of the workshop.

APPENDIX B

WORKSHOP AGENDA

A Workshop on Developing Norms for the Provision of Laboratories in Low-Resource Contexts

Location: The Park Hotel, Stadhouderskade 25
1071 ZD Amsterdam | The Netherlands |

DAY ONE: Wednesday, 27 June 2018

9:00 AM	**Welcome and Goals of the Meeting** Ann Arvin, Chair of the Organizing Committee, Stanford University
9:10	**Overview of the 2011 Istanbul Workshop, Biosecurity Challenges of the Global Expansion of High-Containment Biological Laboratories** Fran Sharples, the U.S. National Academies of Sciences, Engineering, and Medicine (the National Academies)
9:25	**Session 1: What Are Donors Trying to Accomplish?** Session Chair: Ann Arvin What are donors' goals in providing support for biological laboratories in low-resource countries? What level of lab (BSL-2+, 3, other?) are they supporting? What needs are they trying to meet/capabilities they are trying to build? How do they evaluate whether they achieve those goals? Discussion

An Introduction to the Current Picture for High Containment Labs

10:15	**Session 2: The Need for Containment Laboratories: An Overview**

Tom Ksiazek, University of Texas Medical Branch, Galveston National Laboratory

- Ensuring Safe Handling of Pathogens with Pandemic Potential (PPPs) and Other Infectious Agents in Research and Medicine
- Providing an Ability to Make Accurate and Rapid Diagnoses for PPPs and Other Infectious Agents to Ensure Appropriate Medical Care and Outbreak Containment
- Detecting Antimicrobial Resistance in Infectious Agents
- Disease Surveillance and Facilitating Epidemiological Investigations of Infectious Disease Outbreaks
- Detecting Biological Attack Agents and Distinguishing Engineered Agents from Natural Infections

Discussion

10:45 **BREAK**

11:15 **Session 3: The Current Picture for Biological Labs—Estimate of Numbers in Low-Resource Countries: Are They Functioning as Planned?**

Fran Sharples, the National Academies

11:25 **Session 4: Who Is Funding What Where? A Discussion of Building a Map of Projects**

Session Chair: David Harper, Chatham House

- Projects of National Government Donors: Canada, Denmark, France, Germany, Japan, South Korea, Netherlands, UK, USA
- Projects of International Organizations: EU, World Bank, WHO, OIE, FAO
- Projects of Non-governmental Organization/Foundation Donors and Institutions (Mérieux, Pasteur)
- What Others Are Key Players, Are They local or "Imported" and How Are They Involved?—Architects/Designers, Construction Contractors, Equipment Manufacturers, Reagent and Other Suppliers, Inspectors, Biosafety Associations

- Current Distribution of Funds Supporting Biological Labs by All of the Key Players
- What Criteria Do Donors Use for Deciding What to Fund?

12:30 PM **BREAK FOR LUNCH**

12:45 **Luncheon Speaker: Existing International Standards and Codes**
Kazunobu Kojima, World Health Organization

Digging Deeper into What Factors Determine Success

1:45 **Session 5: Key Factors for Building and Sustainably Operating High Containment Labs in Low-Resource Contexts: An Overview**
Nancy Connell, Rutgers University

- Appropriate Infrastructural Components
- Economic/Political Landscape, Low-Resource vs. Ultra Low-Resource Settings
- Adequate Budgets for Power, Water, Equipment, Reagents, and Maintenance Services
- Threats to Safety and Security
- Safety and Security Mechanisms
- Effective Regular Inspections
- Management and Administrative Controls and Culture
- Regulatory Framework, Standards, and Enforcement Mechanisms
- Biosafety and Biosecurity Curricula, Training, and Ensuring Professional Competency

2:00 **BREAKOUT Session 1: Group Discussion of Factors to Consider When Deciding What Biological Laboratory Capabilities a Low-Resource Country Requires**
Breakout Session Chairs: Ann Arvin, David Franz (USAMRIID, ret.)

- Assessment of Site Specific Challenges and Needs
- Obtaining Commitments of Support from In-Country Government Officials

- Soundness and Availability of Necessary Infrastructure: Power, Water, Transportation Network
- Presence or Absence of Country- or Region-Specific Regulatory Framework, Guidelines, and/or Standards
- Availability of Appropriately Trained and Credentialed Local Workforce
- Access to a National or Regional Biosafety Organization
- Biosecurity and nonproliferation considerations

3:30	**BREAK**
4:00	**BREAKOUT Session 1 continued**
5:00	**Report from Chairs of the Breakout Session Groups, Preview of Topics to Be Addressed in Day 2**
5:30	**Adjourn Sessions**
6:00	**Reception Discussions of the Workshop and Needs for Day 2**
7:30	**Adjourn for the day**

DAY TWO: Thursday, 28 June 2018

9:00 AM	**Alternatives to Culture Work *in loco*: Models of Successful Provision of Lower-Containment Laboratory Capacity That Meet Diagnostic and Research Needs** Session Chair: Charles Chiu, University of California at San Francisco Molecular diagnostics: state of the art, readiness, potential future developments, steps, timeline, and roles; work with inactivated pathogens; remote analysis; centralized or regional laboratories—*in loco*, in-country, neighbor-nation, international A review of recent experience in West Africa and practical considerations from working in the field. Jonathan Towner, U.S. Centers for Disease Control and Prevention
10:30	**BREAK**
11:00	**Developing a Candidate Set of Norms**

	Instructions to breakout groups, and sharing of documents on existing guidance and suggestions in development
	BREAKOUT Session 2: Key Dilemmas and Options, Merits, and Downsides to Address Them. Use results to Begin to Develop Candidate Set of Guiding Principles and Common Norms.
	Session Chairs: Ann Arvin, David Franz
12:30 PM	**LUNCH DISCUSSIONS of Morning Session and Goals for Afternoon**
1:30	**Resume Breakout Session 2: Continuation, Revision of Candidate Norms**
3:00	**BREAK**
3:30	**Report from Chairs of the Breakout Session Groups**
	Discussion of Group Candidate Norms, Next Steps
5:00	**ADJOURN WORKSHOP**

APPENDIX C

WORKSHOP PARTICIPANTS AND CONTRIBUTORS

Ann M. Arvin, MD, is the Lucile Salter Packard Professor of Pediatrics and Professor of Microbiology and Immunology, Stanford University School of Medicine, and the Vice Provost and Dean of Research, Stanford University. As Vice Provost, she oversees Stanford's 18 interdisciplinary institutes as well as university research policies, compliance with regulations concerning the responsible conduct of research including human and animal research, and the Office of Technology Licensing. Her laboratory research focuses on molecular mechanisms of varicella zoster virus (VZV) infection and immune responses to this common human herpesvirus. Her clinical research seeks to improve the understanding of the developing immune system in infants and young children in the context of viral infections and vaccines. Her work has been recognized by election to the American Academy of Arts & Sciences, the National Academy of Medicine, the American Association for the Advancement of Science, the Association of American Physicians and the American Pediatric Society. Her past and current national committee service includes the National Academies of Sciences, Engineering and Medicine Board on Life Sciences, the Director's Advisory Council of the National Institute of Allergy and Infectious Diseases, and National Academy of Sciences/National Research Council Committees including the Committee on Federal Research Regulations and Reporting Requirements, the Committee on Policy and Global Affairs, the Committee on Science, Technology and Law, and the Committee on Responsible Science. Dr. Arvin was chief of the Infectious Diseases Division, the Lucile Packard Children's Hospital at Stanford from 1984 to 2006. She received her AB from Brown University, MA in philosophy from Brandeis University, and MD from the University of Pennsylvania. She completed her residency in pediatrics at the University of California, San Francisco (UCSF), and subspecialty training in infectious diseases at UCSF and Stanford University.

Christophe Batejat, MSc, is the deputy director of the laboratory for urgent response to biological threats at Institut Pasteur, Paris. He has been working for more than 15 years on influenza and other emerging viruses as well as biothreats agents, in the areas of laboratory diagnosis, field missions, training, and applied research. His field of research is the survival of viruses outside their host, with a focus on the survival of influenza viruses in aerosols, on surfaces and in water, under different climatic conditions. He runs two biosafety level 3 (BSL-3) laboratories, one of which is being dedicated to the study of bioaerosols, and he is in charge of BSL-3 trainings in Paris. He has been part of many biosafety programs with on-site evaluations and capacity building in the Institut Pasteur International Network. He is involved in the Global Health Security Action Group Lab network, the Global Outbreak Alert and Response Network, and the European EVD-Labnet project.

Sabrina Brizee, MSc, is an international Biosafety and Biosecurity Project Officer, at the Department of Environment of the Centre for Zoonosis and Environmental Microbiology at the National Institute for Public Health and the Environment (RIVM). She participated in various (inter)national projects in the area of biosafety and biosecurity, for example, for the Centre of Excellence and the Dutch Ministry of Foreign Affairs. She assisted in developing and observing national chemical, biological, radiological and nuclear defense (CBRN) field- and table-top exercises in the EU CBRN (Council of Europe (CoE) project 44, of which the main aim was to strengthen CBRN first response in Southeast and Eastern Europe. In addition, she conducted several biosafety and biosecurity capacity-building activities in the East African region, such as implementing the "National Inventory of Dangerous Pathogens." Furthermore, in collaboration with colleagues from Public Health England, she trained 24 participants in Central Asia to become trainers in biosafety and biosecurity, which was part of CoE project 53. At the national level, she worked on a literature study concerning oncolytic viruses and the potential health risk for farm animals, but also developed a national biological field exercise for first responders aimed to enhance forensic investigation/ procedures at CBRN contaminated incident scenes. She has recently been International Federation of Biosecurity Association (IFBA) certified in "Bio-risk Management," by which she required competencies in the fundamental principles and practices of biorisk management, but also successfully completed the BLS-3 course that was provided at the RIVM.

Giovanni Cattoli, DVM, PhD, is the Head of the Animal Production and Health Laboratory in the Joint Food and Agriculture Organization (FAO)/ International Atomic Energy Agency (IAEA) Division of Nuclear Techniques in Food and Agriculture. He has been previously positioned in Italy as Director of the Department of Research & Innovation and the Department of Virology of IZSVe (Istituto Zooprofilattico Sperimentale delle Venezie), which include the FAO and World Organisation for Animal Health (OIE) Reference Laboratory for Newcastle Disease and Avian Influenza, the FAO Reference Centre for Rabies, the OIE Reference Laboratory on fish betanodavirus, and the OIE Collaborating Centre for Diseases at the Human-Animal Interface. In his current position, Cattoli is leading groups working on nuclear and nuclear-derived applications for the development, validation, and application of rapid and innovative diagnostic methods for animal and zoonotic infectious diseases, the development of new vaccines and vaccination strategies for animal infectious diseases, research on pathogenesis, immunology and molecular epidemiology of animal and zoonotic pathogens, animal genetics to improve livestock productions, and disease resistance. He is directly involved in managing several international capacity building and technology transfer activities. He is author or co-author of more than 300 publications including peer-reviewed manuscripts, book chapters, proceeding of conferences and abstracts.

Charles Chiu, MD, PhD, is an associate professor in Laboratory Medicine and Medicine, Infectious Diseases at the University of California, San Francisco (UCSF). He is also the director of UCSF-Abbott Viral Diagnostics and Discovery Center and associate director of the UCSF Clinical Microbiology Laboratory. Chiu heads a translational research laboratory engaged in next-generation sequencing approaches for diagnosis of infectious diseases, pathogen discovery, bioinformatics software development analysis, nanopore sequencing, and characterization of emerging infections (Lyme disease, enterovirus D68, Ebola virus, and Zika virus). His work is supported by research grants from the National Institutes of Health, Abbott Laboratories, bioMerieux, Global Lyme Alliance, philanthropic organizations (Sandler, Bowes, Marcus, Charles and Helen Schwab, and Steve and Alexandra Cohen Foundations), and the California Initiative to Advance Precision Medicine. Chiu collaborates with partners around the world to sequence pathogens from emergent infectious disease outbreaks. He previously served on the National Academies Committee on Polymerase Chain Reaction Standards for the BioWatch Program. Chiu obtained an MD and PhD in biophysics from the University of California, Los Angeles, and subsequently completed an residency, fellowship, and postdoctoral research at UCSF. He has authored more than 80 peer-reviewed publications, holds more than 15 patents and patent applications, and serves on the scientific advisory boards for Therabio, Inc.

John Paul Clark is an epidemiologist and health planner. He is the coordinator of the World Bank's Regional Disease Surveillance Systems Enhancement Program and provides technical leadership in the areas of maternal and child health and communicable diseases. Prior to joining the World Bank in 2006, Clark held senior positions at the World Health Organization, the U.S. Department of Health and Human Services, and USAID. Over the past 30 years, he has made numerous contributions to public health research, policy, and practice and has led innovative multi-sector and cross-border efforts to reduce poverty and foster economic development through disease prevention, control and elimination. Clark holds advanced degrees from the Johns Hopkins University and the London School of Economics and Political Science.

Nancy D. Connell, PhD, is a professor and director in the Division of Infectious Disease in the Department of Medicine at the University of Medicine and Dentistry of New Jersey (UMDNJ), New Jersey Medical School. A Harvard University PhD in microbiology, Connell's major research focus is the interaction between respiratory infectious agents, such as M. tuberculosis and B. anthracis, and the macrophage. She is director of the biosafety level 3 (BSL-3) facility of UMDNJ's Center for the Study of Emerging and Re-emerging Pathogens and chairs the University's Institutional Biosafety Committee. Connell has served on or chaired numerous NIH review panels. She has served on more than 15 committees of the National Academy of Sciences, Engineering, and Medicine, including the Committee to Review the Scientific Approaches used in the FBI's Investigation of the 2001 *Bacillus anthracis* Mailings (2011). Dr. Connell chairs the National Academies Standing Committee for Faculty Development for Education about Research with Dual Use Issues in the Context of Responsible Science and Research Integrity, which has directed sustainable training workshops held across the Middle East and North Africa. Connell was recently appointed to the National Academies Board on Life Sciences and received the 2017 Outstanding Scientist Award from the Edward J. Ill, M.D. Excellence in Medicine Awards.

David R. Franz, DVM, PhD, served in the U.S. Army Medical Research and Materiel Command for 23 of 27 years on active duty and retired as colonel. He served as Commander of the U.S. Army Medical Research Institute of Infectious Diseases and as Deputy Commander of the Medical Research and Materiel Command. Prior to joining the Command, he served as group veterinarian for the 10th Special Forces Group (Airborne). Franz was the chief inspector on three United Nations Special Commission biological warfare inspection missions to Iraq and served as technical advisor on long-term monitoring. He also served as a member of the first two U.S.-U.K. teams that visited Russia in support of the Trilateral Joint Statement on Biological Weapons and as a member of the Trilateral Experts' Committee for biological weapons negotiations. He is a member of the National Academy of Sciences Committee on International Security and Arms Control. He previously served on the Board on Life Sciences and the Department of Health and Human Services National Science Advisory Board for Biosecurity. Franz also co-chaired the National Academies' Committee on Strengthening and Expanding the Department of Defense Cooperative Threat Reduction Program. He serves on the Board of Integrated Nano-Technologies, LLC. Franz holds a DVM from Kansas State University and a PhD in physiology from Baylor College of Medicine.

David Harper, PhD, is the Managing Director of Harper Public Health Consulting Limited. He is also Senior Consulting Fellow at the Chatham House Centre on Global Health Security. Previously, Harper was Special Adviser to the Assistant Director-General for Health Security and Environment at the World Health Organization in Geneva, where his principal role was to advise on Global Preparedness for Health Security. Before March 2012, Harper was the Chief Scientist and Director General for Health Improvement and Protection in the UK Department of Health. He was responsible for protecting the population from risks posed by infectious diseases and environmental hazards; preparing for, and responding to, a range of health emergencies and disruptive challenges to health services; reducing the burden of conditions associated with poor lifestyles; and promoting health and wellbeing. He also held the international health and scientific development portfolios for the Department of Health. A scientist by training, Harper graduated in microbiology from the University of Dundee and gained his PhD in biochemistry from the University of Birmingham. He is a Fellow of the Royal Society of Biology, a Fellow of the Faculty of Public Health of the Royal College of Physicians, and an honorary Fellow of the Royal Society of Public Health. He was awarded the Commander of the Order of the British Empire in 2002. He has honorary professorships at the London School of Hygiene and Tropical Medicine and the University of Dundee, and an honorary Doctorate of Science degree from Cranfield University, where he is also a visiting professor.

Andrew Hollands has been involved in the Defense Threat Reduction Agency, Cooperative Threat Reduction, Biological Threat Reduction Program (BTRP) for the past 6 years, working on the Southeast Asia and Africa portfolios. Hollands has led efforts to strengthen laboratory and field biosurveillance capacity across both regions. In addition, Hollands led training and policy development efforts to elevate country capabilities in biosafety and biosecurity for better alignment with international guidelines and best practices. Hollands is currently the Africa Region Lead responsible for engaging with countries for biosecurity, biosurveillance, and biosafety capacity development and developing the program's engagement strategy for the continent. Prior to BTRP, Hollands helped to manage the Department of Homeland Security, Science and Technology, Chemical and Biological Defense research and development programs for national defense.

Mitsuo Isono, MD, PhD, obtained MD and PhD degrees in clinical medicine from Tohoku University School of Medicine. After engagement in clinical medicine in Oita Medical University, he started work as a senior technical advisor for the health sector in Japan International Cooperation Agency, the implementing organization for the official development assistance by the Government of Japan, since 2007. The main areas of works are infectious disease control, noncommunicable disease control, and construction of hospitals and laboratories. While working as TB program advisor in Afghanistan, he has engaged missions for those areas in about 40 countries in Asia, Middle East, Africa, and Central America.

Barbara Johnson, PhD, owns the consulting company Biosafety Biosecurity International. She is a microbiologist with more than 25 years of experience in the U.S. government and private industry in the areas of biosafety, biocontainment, and biosecurity. Currently, she develops site-specific risk assessments and mitigation strategies, assists in developing frameworks internationally to establish Institutional Biosafety Committees and support programs, reviews and develops designs for biocontainment facilities (A/BSL-2 through A/BSL-4 and BSL-3 Ag), certifies and validates containment laboratories, develops and provides biosafety and biosecurity training in the United States and internationally (more than 20 countries), and provides strategic and technical assistance in developing national-level and international biosafety, biosecurity, and biorisk management programs for conducting work with high-risk pathogens. She has served on several panels for the National Academies. Johnson is a Registered Biosafety Professional, approved BSL-3 Facility Certifier and Trainer by the Singapore Ministry of Health, past president of ABSA, Co-Editor-in-Chief of the American Biological Safety Association journal *Applied Biosafety*, past vice president of A-PBA, founding member of IFBA, and a past President of the American Biological Safety Association.

Kazunobu Kojima, PhD, graduated from Hokkaido University School of Medicine in Sapporo, Japan. He subsequently obtained a PhD from Sapporo Medical University for his study in infectious disease epidemiology. He was an Assistant Professor at its medical school, having studied and taught virology and epidemiology with particular research interests in rotavirus and poliovirus, including long-term field experience in Myanmar engaged in polio eradication initiative. Kojima has been in service to the World Health Organization for more than 14 years, starting from the Regional laboratory coordinator at WHO Western Pacific Regional Office (WHO/WPRO) in Manila, the Philippines. He moved to the WHO Lyon Office and then to Headquarters in Geneva in 2010, where he continues as the scientist charged with the responsibility for biosafety and laboratory biosecurity, including transportation of infectious substances.

Thomas G. Ksiazek, DVM, is director of high-containment laboratory operations at the Galveston National Laboratory and is a virus expert with 40 years of experience on the front lines of infectious disease research. Through the years he has worked on disease discovery and outbreak response efforts in Asia, Africa, South America, and the Middle East. Ksiazek is an expert in hemorrhagic fevers, such as Ebola, and viral diseases. He is credited as being one of the co-discoverers of SARS (Severe Acute Respiratory Syndrome), which appeared in China in 2002. His quick work in identifying the virus is often credited as one reason why the disease was contained quickly. He is a three-time recipient of the Secretary of Health and Human Services Award. He received a Lifetime Achievement Award for Filovirus Science at the 6th International Filovirus Symposium in 2014, and in June 2015 he was named Distinguished Alumnus of the Kansas State College of Veterinary

Medicine. In addition to serving as a senior staff scientist and director of high-containment operations, Ksiazek is a professor in the departments of Pathology and Microbiology & Immunology at the University of Texas Medical Branch in Galveston. He served as the Chief of the Special Pathogens Branch at the Centers for Disease Control and Prevention since 1991 after retiring from the U.S. Army as Lieutenant Colonel with 20 years of active duty service. Ksiazek earned his DVM in 1970, and then spent a year as associate veterinarian at the Adirondack Animal Hospital in Glensfalls, New York. He started his military career when he joined the U.S. Air force in 1971, holding a position that year as base veterinarian at Sheppard Air Force Base, Texas. He then worked as chief of Veterinary Services, Royal Air Force, Chicksands, U.K.

Micah Lowenthal, PhD, is senior director for international networks in Policy and Global Affairs of the National Academies. He conducts and oversees a variety of international engagements and studies on nuclear, biological, space, and cyber safety and security. Previously, Lowenthal was a lecturer and researcher at the University of California (UC), Berkeley. He is an elected fellow of the American Association for the Advancement of Science (AAAS) and the American Physical Society (APS), and a past chair of the APS Forum on Physics and Society. Lowenthal earned an AB in physics and a PhD in nuclear engineering, both from UC Berkeley.

Craig Reed, PhD, received his BS in molecular biology from Vanderbilt University and a PhD in biochemistry and molecular biology from the University of Georgia. Following his doctoral work, he served as a Captain in the U.S. Army and was stationed at the U.S. Army Medical Research Institute of Infectious Diseases. Reed is founder and chief executive officer of Inspirion Biosciences, an international consulting firm specializing in containment laboratory bio-risk management. Reed has supported the Departments of Defense, State, and Health and Human Services. He has worked closely with the directors and staff of more than 50 epidemiological surveillance labs and biological research facilities in more than 25 different countries to improve containment laboratory infrastructure, laboratory work practices and administrative controls, and other aspects of biosafety and biosecurity. He led and supported two European Committee for Normalization Workshops on laboratory bio-risk management. Reed is a Registered Biosafety Professional. He is an active member of the biosafety associations of Brazil, Africa, Europe, and the Asia-Pacific region, and served as President of the Chesapeake Branch of the American Biological Safety Association. He has served as an advisor to the International Federation of Biosafety Associations and the Biosafety Association of Central Asia and the Caucasus.

Masayuki Saijo, MD, PhD, is Director of the Department of Virology at the National Institute of Infectious Diseases in Tokyo, Japan. Dr. Masayuki Saijo obtained his MD and his PhD in pediatrics with Professor Yoshioka and Professor Okuno, Asahikawa Medical University, in Japan in 1991. He has studied clinical infectious diseases such as RSV infections in children and antiviral-resistant herpes virus infections in immunocompromised subjects. He joined the National Institute of Infectious Diseases, Tokyo, Japan, in 1997, and has studied viral hemorrhagic fevers such as Ebola, Marburg, and Crimean-Congo hemorrhagic fevers. His research team is leading in the field of diagnostics and clinical aspects of emerging virus infections including viral hemorrhagic fevers.

Frances E. Sharples, PhD, is the Director of the National Academies' Board on Life Sciences (BLS). She is responsible for the management of all BLS projects, maintaining their quality, tracking their budgets and deliverable milestones, and making decisions relevant to staffing of projects. BLS serves as the National Academies' focal point for a wide range of technical and policy topics in the life sciences, including genomics, biodiversity conservation, bioterrorism, and key topics in basic research, such as gene editing. Immediately prior to joining the National Academies, Sharples was a Senior Policy Analyst for the Environment Division of the Clinton Administration's White House Office of Science and Technology Policy (OSTP) from October 1996 to October 2000. Sharples went to OSTP from the U.S. Department of Energy's Oak Ridge National Laboratory, where she served in various positions in research and management in the Environmental Sciences Division between 1978 and 1996. Sharples received her BA in biology from Barnard College (1972) and her MA (1974) and PhD (1978) in zoology from the University of California, Davis. She served as an American Association for the Advancement of Science (AAAS) Environmental Science and Engineering Fellow at the Environmental Protection Agency during the summer of 1981, and served as an AAAS Congressional Science and Engineering Fellow in the office of Senator Al Gore in 1984-1985. She was a member of the National Institutes of Health's Recombinant DNA Advisory Committee in the mid-1980s, and was elected a Fellow of the AAAS in 1992.

Jonathan Towner, PhD, leads the Virus Host Ecology Section within the Viral Special Pathogens Branch at the Centers for Disease Control and Prevention (CDC). His team focuses on the ecology of high-consequence bat-borne viruses including filoviruses (e.g., ebolaviruses and marburgviruses) and paramyxoviruses with emphasis on (1) identifying their natural reservoir hosts, (2) determining the mechanisms used by these viruses to persist long term in nature, and (3) identifying the drivers that cause virus spillover to humans. In addition to his research, as part of a major public health agency, Towner responds when needed to viral hemorrhagic fever outbreaks in Africa to establish and/or execute on-site molecular diagnostic testing. In this capacity, he has helped establish or operate field labs at four major filovirus outbreaks since 2000, including the CDC lab in Bo, Sierra Leone, that processed more than 27,000 human diagnostic specimens. He has been well trained in filovirus biology and ecology by leading authorities in the field including Drs. Stuart Nichol, Thomas Ksiazek, Robert Swanepoel, and Pierre Rollin. Towner received both his BA and PhD at the University of California, Berkeley, and has more than 25 years of training as a molecular virologist and 19 years of experience conducting virus research under BSL-4 containment.

Sapana Vora, PhD, joined the U.S. Department of State's Biosecurity Engagement Program (BEP) in the Office of Cooperative Threat Reduction (CTR) as an AAAS Science and Technology Policy Fellow in 2015. BEP's mission is to reduce the threat of bioterrorism by preventing terrorist access to potentially dangerous biological materials, dual-use technology, and bioscience expertise. As BEP's Deputy Team Chief,. Vora oversees BEP's annual funding cycle, helps shape CTR programmatic and policy strategies and projects, and participates in a number of interagency science policy discussions, including those on biological select agents and toxins, global health security, and other biosecurity issues. Prior to joining BEP, she was a Mirzayan Science and Technology Policy fellow and research associate at the National Academies, where she worked on the "Ovarian Cancers: Evolving Paradigms in Research and Care" consensus study. She holds a PhD in cancer biology from the University of Chicago and a BS in biology and English from the University of North Carolina at Chapel Hill.

APPENDIX D

WORLD BANK 2017 LIST OF LOW- AND LOWER-MIDDLE INCOME ECONOMIES

Low-Income Economies ($1,025 or less)

Afghanistan
Benin
Burkina Faso
Burundi
Central African Republic
Chad
Comoros
Congo, Dem. Rep.
Eritrea
Ethiopia
Gambia, The
Guinea
Guinea-Bissau
Haiti
Korea, Dem. People's Rep.
Liberia
Madagascar
Malawi
Mali
Mozambique
Nepal
Niger
Rwanda
Senegal
Sierra Leone
Somalia
South Sudan
Tanzania
Togo
Uganda
Zimbabwe

Lower-Middle-Income Economies ($1,026 to $4,035)

Armenia
Bangladesh
Bhutan
Bolivia
Cabo Verde
Cambodia
Cameroon
Congo, Rep.
Côte d'Ivoire
Djibouti
Egypt, Arab Rep.
El Salvador
Ghana
Guatemala
Honduras
India
Indonesia
Kenya
Kiribati
Kosovo
Kyrgyz Republic
Lao PDR
Lesotho
Mauritania
Micronesia, Fed. Sts.
Moldova
Mongolia
Morocco
Myanmar
Nicaragua
Nigeria
Pakistan
Papua New Guinea
Philippines
Samoa
São Tomé and Principe
Solomon Islands
Sri Lanka
Sudan
Swaziland
Syrian Arab Republic
Tajikistan
Timor-Leste
Tonga
Tunisia
Ukraine
Uzbekistan
Vanuatu
Vietnam
West Bank and Gaza
Yemen, Rep.
Zambia

APPENDIX E

LIST OF LABS IDENTIFIED IN LOW-RESOURCE COUNTRIES

Country	# of BSL-2/3/4[a]	Type of Lab[b]	Funder[c]	Status[d]	Capability
Afghanistan	BSL-2, no known BSAT	CPHL		Unknown	PCR, ELISA, microbiology
Armenia	BSL-2?, no known BSAT	MDC (Govt/Military?, Academic)	INTAS (India 2006), ISTC (2014) CRDF (2011)	Operational	General and applied research, maintains strain collection
Bangladesh	(2) BSL-2, Modular Work with Nipah for US CDC; works with B. anthracis human outbreaks	One locally built) (Min Health) IEDCR (one modular from TechComp—China	US CDC	Locally built is operational Modular likely in use but BAS-HVAC mal-functions	Microbiology, serology, ELISA
	BSL-2 No known BSAT	icddr,b (nonprofit, Health)	Gates Foundation, DFID (UK), DFATD (CA), Sida (SWE), USAID, others	BSL-2 operational, BSL-3 decommissioned 2016; new BSL-3 in construction	Microbiology, serology, ELISA; PCR, sequencing (Nipah, CCHF) TB, influenza, NIPAH for US CDC, V. cholera
	BSL-2 possible BSAT	BLRI, Min Ag	OIE	Operational	ELISA, PCR, DNA sequencing, vaccine

Country	# of BSL-2/3/4[a]	Type of Lab[b]	Funder[c]	Status[d]	Capability
		(veterinary reference lab)			development, basic research
	BSL-2 Isolates BSAT- B. abortus, Brucella ssp	CDIL Min Ag (veterinary reference lab)	USAID, OIE	Operational	PCR, serology, histopathology ELISA, HPAI preparedness
Burkina Faso	BSL-3 Container Lab	Centre Muraz	Several sources	Operational	Arbovirus (yellow fever, dengue, chikungunya, zika) and VHF (RVF, CCHF) research and diagnostics; serology/PCR; limited sequencing capacity
Cambodia	BSL-3 Work with BSAT, XDR TB, emerging influenza	IP, C Min Health	Funds: INSERM, ANRS, IRD, Fondation Mérieux and Institut Pasteur WHO Global Influenza Surveillance Center	Operational	Research and diagnostics: virology, bacteriology, FACS, gene sequencer, multiplex assays
	BSL-2 BSAT unknown	NaVRI	OIE	Operational	Diagnostics, HPAI, unknown capability

Country	# of BSL-2/3/4[a]	Type of Lab[b]	Funder[c]	Status[d]	Capability
		Veterinary Reference Lab			
Cameroon	BSL-3 No known BSAT	PC, C Min Health,	Institut Pasteur (?), WHO National Reference Center for Avian Flu	Operational	Diagnostics, epidemiology, research on local disease
Congo (Dem. Rep)	BSL-3 BSAT include Ebola, Marburg	INRB Min Health, (WHO Natl Ref Lab for measles and other disease)	Presidents Malaria Initiative (USA) 2016 Japan International Cooperation Agency (JICA) 2017	Operational	Human diagnostics, ELISA, research
Cote d'Ivoire	BSL-3 BSAT (Lassa Fever Virus)	IP, Cd'I Commercial	Unknown	Operational	Regional AI reference lab, human diagnostic tests
Djibouti	BSL-3 No known BSAT	Natl TB Lab Min Health,	AFD (Agence Française de Développement) 2010	Operational	Located at Hôpital Paul Faure—human TB diagnosis
Egypt	BSL? 2 BSAT	MIRCEN Cairo, Academic	UNESCO; Ain Shams University	Operational	Collecting, identifying and preserving microbial strains located at Ain

Country	# of BSL-2/3/4[a]	Type of Lab[b]	Funder[c]	Status[d]	Capability
	BSL-3 No known BSAT	U.S. Navy Medical Research Unit-3; Govt/MIL	Department of Defense Global Emerging Infections Surveillance and Response System, 2006	Operational (no longer operational?)	Shams University. Surveillance sample testing mainly on RT-PCR methodology
Ethiopia	BSL-3 Possesses agricultural BSAT	National Animal Health Diagnostic and Investigation Center; Min Health	?	Operational	National reference laboratory, coordinates national surveillance programs for transborder animal diseases
	BSL-3 (mobile?) Crossover and animal BSAT	Pan African Veterinary Vaccine Centre of the African Union; Academic	?	Operational	Produce and distribute animal vaccines
	BSL-3 No BSAT	Armauer Hansen Research Institute; Min Health	Swedish International Development Cooperation Agency (current); Norwegian Agency for Development	Operational	Biomedical research, basic (immunology and molecular biology), epidemiological, and translational research

Country	# of BSL-2/3/4[a]	Type of Lab[b]	Funder[c]	Status[d]	Capability
			Cooperation (current); Ethiopian Government (current)		
	BSL-3 No BSAT	Ethiopian Public Health Institute; Min Health	World Bank, 2015 ($49M)	Operational	Many US, NGO, and other country collaborations
	BSL-3 No BSAT	TB Reference Laboratory	PEPFAR, Global Fund to Fight AIDS, TB and Malaria, USAID Tuberculosis Control Assistance Program (?)	Operational	At St. Peter's Hospital-advanced diagnostics for TB and MTB
Georgia	BSL-3 BSAT-ASF virus and CCHF virus	Central Public Health Reference Laboratory; Min Health, Min Ag	USG—$100M in 2011	Operational	Human and animal disease surveillance and research
	8 BSL-3—no info				

Country	# of BSL-2/3/4[a]	Type of Lab[b]	Funder[c]	Status[d]	Capability
Ghana	BSL-3 BSAT- Nipah virus	Kumasi Center for Collaborative Research in Tropical Medicine, Academic	Unknown	Operational	Reference Lab/Culture Collection, research on tropical diseases
	BSL-3 No BSAT	Department of Virology, Noguchi Memorial Institute for Medical Research, Academic	25+ funding sources listed with no date or confirming info	Operational	Reference Lab/Culture Collection, antiviral research, molecular epidemiology of polioviruses, enteroviruses, HIV, and others
	3 other BSL-3, 2 with BSAT, B. anthracis	Located in Accra, Takoradi, and Tamale	Global Affairs Canada GPP program	Operational	Government veterinary services labs, human and animal disease surveillance (B. anthracis and avian influenza)

Country	# of BSL-2/3/4[a]	Type of Lab[b]	Funder[c]	Status[d]	Capability
Guatemala	BSL-3 1 BSAT- B. abortus	National Health Laboratory; Min Health	?	Operational	Reference Lab/Culture Collection
Guinea	BSL-3 2BSAT- EBOV, Lassa	Virology Lab, Department of Infectious and Tropical Diseases; Academic	Howard Hughes Medical Institute, (?); European Union (?)	Operational	At Donka Hospital; viral hemorrhagic fever research, many outside collaborators listed (no date)
	BSL-3	Institute of Microbiology, University of Conakry; Academic	Several EU and corporate funders listed (no date/$)	Operational	Best-equipped lab for VHF and YF diagnostics in this part of West Africa
Haiti	BSL-3 3 BSAT: Y. pestis, C. botulinum, B. anthracis	Tuberculosis Laboratory, University of Florida Public Health Research Center; Haiti; Academic	Armed Forces Health Surveillance Center (US); NIH (US)	Operational	TB, cholera, and viral disease epidemiology and diagnostics
	BSL-3 No BSAT	GHESKIO; Min Health	4 funders listed (no data), includes	Operational	HIV, STD, and TB diagnostics, treatment, and research

Country	# of BSL-2/3/4[a]	Type of Lab[b]	Funder[c]	Status[d]	Capability
		(2 campuses with another? BSL-3 complete 2010)	USAID and UNITAID		
	BSL-3 No BSAT	National Public Health Laboratory; Min Health	Numerous engagers; Funders: US CDC Foundation; GE Foundation (US); Kaiser Permanente; Robert Wood Johnson Foundation	Operational	Virology, bacteriology, immunology, epidemiology, has US CDC onsite
India	44 BSL-3 collected—Many with BSAT	Mix of private vaccine/drug manufacturers, MIL, academic, hospital labs, human and animal	In general, no data on funders	Operational	Virology, bacteriology, diagnostics, vaccine production, epidemiology, etc.
	BSL-4	National Institute of Virology; Pune (Min Health, possible MIL)	WHO Arbovirus Collaborating Center	Operational	'Biodefense lab', virus epidemiology (Influenza, Nipah, CCHF, KFD, SARS, etc.
Indonesia	22 listed as BSL-3 (3 seen—confirm)	MIL, academic, hospital labs, Min Health, Min Ag	Funded by Indonesia Govt; Vet labs have received significant	22 operational 10 unknown	Diagnose human and animal virus, bacteria, parasite infection (rabies

Country	# of BSL-2/3/4[a]	Type of Lab[b]	Funder[c]	Status[d]	Capability
	7 w/ BSAT, another 10 unknown, 8 RVDLs not listed (BSL-2); 2 RVDL have BSL-3 donated by Japan		contributions of equipment, reagents, PPE and i.e. BSC certification—repair from OIE, FAO, JICA for HPAI and some USAID training	8 vet labs operational at BSL-2; 2 Japanese donated BSL-3 container labs at RVDL are not functional (no funding for power; airflow not working)	is a big problem), epidemiology, HIV research, etc.
Kenya	9 BSL-3 7 listed as possessing BSAT	Min Health, Min Ag, MIL	Generally no discussion but mention of Wellcome Trust, U.S. Army Medical Research Unit-Kenya	9 operational, 3 in construction or planning	Diagnosis and research on human and animal virus, bacteria, infection; collection and repository; vaccine–drug production
Kyrgyz Republic	BSL-3 (modular-German built) No BSAT	National Reference Laboratory; Min Health	German Development Bank (2013); Global Fund to Fight Aids, Tuberculosis and Malaria (2003-2009)	Operational	Reference Lab/Culture Collection; TB treatment, diagnostics, prevention
Lao, PDR	BSL-3 1 BSAT- B. pseudomallei	National Tuberculosis	Wellcome Trust, WHO	Operational	At Central Mahosot Hospital; Reference Lab/Culture Collection

Country	# of BSL-2/3/4[a]	Type of Lab[b]	Funder[c]	Status[d]	Capability
		Reference Center; Min Health, MIL			and research on epidemiology, treatment of malaria, typhoid, typhus, and other community acquired pathogens.
	BSL-2 BSAT- Unknown	National Animal Health Center; Min Ag	Staff are affiliated with Oxford Univ & Wellcome Trust	Operational- upgrade status unknown	Diagnosis of animal disease
Liberia	BSL-2 BSAT – EBOV strains	Liberian Institute for Biomedical Research; Min Health, MIL	During EBOV outbreak 2014 received staffing, reagents, equipment from US CDC, USMRIID, USAID, US NIH	Operational	National lab for diagnosis of hep B, malaria, cholera, and in 2014 ebola; limited Lassa fever PCR capacity
Madagascar	BSL-3 2 BSAT- RVF virus and Y. Pestis	Institut Pasteur in Madagascar, Min Health, commercial	Institut Pastuer (Paris)—2017 providing staff and RT-PCR equipment – reagents. 14 funders (most NGOs some	Operational	national reference lab; studies malaria entomology, plague, mycobacteria, experimental bacteriology, immunology

Country	# of BSL-2/3/4[a]	Type of Lab[b]	Funder[c]	Status[d]	Capability
			commercial) listed with no $ or year WHO Collaborating Center for Plague; WHO Reference Laboratory for Polio, Measles and Influenza		
Malawi	BSL-3 No BSAT	Tuberculosis Central Reference Laboratory, USAID?, MIL/Govt	USAID	Operational	Reference Lab/Culture Collection, TB, MDR-TB epidemiology and diagnosis
Mali	BSL-3 No BSAT	FMPOS; Min Health	NIAID, 2002 and 2010 (infrastructure for BSL-3)	Operational	Reference Lab/Culture Collection
Moldova	BSL-3 No BSAT	National Center for Public Health, Min Health	???	Operational	Develops legislative, regulatory public health, health promotion, and noncommunicable disease surveillance
Morocco	6 BSL-3 listed with BSAT (RVFV,	Human and animal health; Min Health, Min Ag	Institut Pasteur in Morocco Casablanca	6 operational	Collection, research, disease diagnosis,

Country	# of BSL-2/3/4[a]	Type of Lab[b]	Funder[c]	Status[d]	Capability
	AHS virus); 3 BSL-3 planned.		is established, no recent funding		veterinary vaccine production
Mozambique	BSL-3 2 BSAT- B. anthracis, Newcastle virus	Agricultural Research Institute of Mozambique, Min Ag	None listed	Operational	Reference Lab/Culture Collection
Myanmar	BSL-3 No BSAT	National Tuberculosis Reference Laboratory; Min Health, UNITAID	UNITAID $87.6M but no date	Operational	Reference Lab/Culture Collection; TB MDR TB diagnosis
	BSL-3 No BSAT	No. 2 Defence Service General Hospital; Govt/MIL?	No info	Operational	Hospital/clinic with full diagnostic capability
	BSL-3 No BSAT	NTRL Regional Laboratory, Min Health, UNITAID, FIND	Lists several NGOs but no date/amount	Operational	TB Reference Lab/Culture Collection
	BSL-2 BSAT unknown	YNRL, Min Ag	BSCs donated 2008 and 2011 OIE and JICA	Operational	Animal health diagnostic lab

Country	# of BSL-2/3/4[a]	Type of Lab[b]	Funder[c]	Status[d]	Capability
	BSL-2 BSAT unknown	MVDL, Min Ag	BSCs donated 2011 OIE and JICA	Operational	Animal health diagnostic lab, bird flu, rabies, etc.
Nepal	BSL-2 BSAT likely	Central Veterinary Laboratory, Min Ag	OIE donated HPAI reagents and equipment (BSC, PCR, RT PCR), PPE and lab equipment (approx. 2010).	Operational Located in Kathmandu	Lab tests for HPAI, rabies, Newcastle, peste des petites ruminants, Infectious bursar disease, bacteria, etc.
Nicaragua	BSL-3 No BSAT	Veterinary Diagnostic Laboratory, Min Ag	Sources from US, EU, El Salvador but no date/amount	Operational	Analysis and diagnosis of veterinary and food products
Nigeria	10 BSL-3 with numerous endemic BSAT (e.g., Ebola, Lassa)	Institutes are academic, Min Ag, Lagos State Min Health, Mainland Hospital	Global Affairs Canada GPP program and numerous other engagements and possible funders	Operational, one just completed	WHO National Influenza Center, WHO reference center for arboviruses; human and animal virus epidemiology, surveillance, isolation, diagnostics, characterization, and research. Capacity to handle unknown pathogens; PCR, sequencing; reference

Country	# of BSL-2/3/4[a]	Type of Lab[b]	Funder[c]	Status[d]	Capability
					lab/culture collection (TB/HIV), biobank
BSL-2	Min Ag, Lagos State Min Health		Global Affairs Canada GPP program	Just completed	Human and animal disease surveillance and diagnostics
Pakistan	5 BSL-3, 2 with BSAT (B. anthracis); 6 BSL-3 unknown	Academic, commercial, medical, Min Ag, Min Health, MIL?	WHO Collaborative Centre for Research and Training in Viral Diagnostics; No info on funding or engagements	5 operational 6 unknown, or in planning/construction	Research on antimicrobial resistance and emerging viral and bacterial infections; culture collection; human and animal clinical diagnosis; TB Reference Laboratory
Philippines	3 BSL-3, 1 w/BSAT (B. melitensis) 6 unknown	Academic, Govt/MIL, Min Health	Funding info is largely blank; Global Fund to Fight AIDS, Tuberculosis and Malaria (Japan, 2011)	3 operational 6 unknown BSL, in construction or not collected	Reference Lab/Culture Collection, human disease diagnosis.
	BSL-2	Philippines Animal Health Center, Min Ag	UNDP/FAO equipment provided 2007-2009	Operational	Veterinary Reference Lab Quezon City
Senegal	BSL-3 BSAT Unknown	Institut Pasteur in Dakar, Min Health	WHO Collaborating Centre for	Operational (2016)	Diagnosis, treatment, and vaccination of

Country	# of BSL-2/3/4[a]	Type of Lab[b]	Funder[c]	Status[d]	Capability
			arboviruses and VHFs; WHO National Lab for flu, measles and polio. Funding from French Govt and EU ERDF program; Unknown funding from Pasteur Inst, WHO		arboviruses, VHF, and enteroviruses; Reference Lab/Culture Collection
Sierra Leone	BSL-3 2 BSAT (EBOV spp, Lassa)	Kenema Government Hospital, Viral Hemorrhagic Fever Laboratory	Significant collaboration with US CDC, USAMRIID, universitites and companies; UK foundations. Recent funding unknown	Operational	Diagnosis and treatment of Lassa fever and other diseases
	BSL-3 No BSAT	Central Public Health Reference Laboratory	Amount/year of funding unknown	Operational	Reference Lab/Culture Collection for HIV and other
	BSL-3 BSAT unknown	Chinese Mobile Container BSL-3 Lab (3 containers)	???	Operational	VHF diagnosis

Country	# of BSL-2/3/4[a]	Type of Lab[b]	Funder[c]	Status[d]	Capability
Sri Lanka	BSL-3 1 BSAT (Y. pestis) 3 other entries w/no data	Medical Research Institute, Colombo, Min Health	None listed	Unclear website states both not operational and operating	Unknown
Syria	BSL-3 1 BSAT (B. abortus)	Atomic Energy Commission of Syria; Govt/MIL	Amount/year of funding unknown	Operational in Damascus	Work on Brucella bacteria
Tajikistan	BSL-3 No BSAT	National Public Health Reference Laboratory; Min Health	Global Fund to Fight Aids, Tuberculosis and Malaria (2013)	Operational	Reference Lab/Culture Collection Molecular and culture testing for viral and bacterial diseases
	BSL-3 No BSAT	National TB Reference Laboratory; Min Health	No funding described	Operational	At Machiton Hospital; Reference Lab/Culture Collection diagnostics of TB, HIV, malaria, etc.
Tanzania	8 BSL-3 2 w/BSAT (RVF virus, Y. pestis, B. anthracis, Rinderpest)	Min Health, Min Ag, Hospital, Commercial,	Amount/year of funding unknown	7 operational 1 planned (has BSAT)	Reference Lab/Culture Collection; Surveillance of Rift Valley fever, H1N1, and dengue, diagnosis of human and animal diseases

Country	# of BSL-2/3/4[a]	Type of Lab[b]	Funder[c]	Status[d]	Capability
Uganda	BSL-2	Makerere University Infectious Disease Institute, Academic	Amount/year of funding unknown	Operational	Reference Lab/Culture Collection, HIV/TB diagnosis
	BSL-3 No BSAT	UVRI; Med Research Council	Amount/year of funding unknown; CDC	Operational	Hospital; HIV/AIDS; VHF program (serology, PCR, limited sequencing)
	6 BSL-3 No BSAT	Min Ag, Min Health	Some funders listed: USAID, Global Fund, UK Medical Research Council, Gates Foundation	Unknown, construction, not collected	Reference lab, hospital, culture collection
Ukraine	4 BSL-3 4 w/ BSAT (B. anthracis, Y. pestis, F. tularensis, B. abortus, Guanarito virus)	(Some part of FSU weapons program). Min Health	No funding described	Operational	Reference Lab/Culture Collection, bacteriologic and arbovirus research, surveillance of 'plague' diseases
Vietnam	BSL-3 No BSAT	National Institute of Hygiene and Epidemiology Min Health	JICA (2008), WHO, The Global Fund	Operational	Reference Lab/Culture Collection, human disease prevention and control

Country	# of BSL-2/3/4[a]	Type of Lab[b]	Funder[c]	Status[d]	Capability
	BSL-3 and SAPO 4 Likely BSAT	Virology Reference Laboratory, Min Health	Wellcome Trust, Canada PHAC & US HHS provide staff	Operational	At Oxford University Clinical Research Unit; diagnosis/research planned on TB, influenza, rabies, FMD, TBE virus, CHIKV, bat corona viruses
	3 others (1 BSL-3 w/ BSAT- Y. pestis)	Min Health, Academia, unknown	Unknown	Unknown	Unknown
	BSL-2 BSAT unknown	Central Veterinary Research Laboratory, Min Ag	OIE	Operational	At Hanoi Agriculture University; reference and diagnostic lab for animal health
West Bank aka Palestine	BSL-3 No BSAT	PCPHL; Min Health (may have MIL ties, work on Rad Protection)	Agence Francaise de Developpement (2011)	Operational	Human diagnostics of highly pathogenic organisms (i.e., TB)
Zambia	6 BSL-3 1 with BSAT (B. anthracis, Y. pestis,	Institutes are academic, veterinary school, and hospital/clinic	Container labs 'managed' by The Zambia AIDS Related	Operational 4 of 6 are modular 40 ft container labs (ZamLab)	Surveillance and diagnosis of hemorrhagic fever viruses, TB, plague bacillus, trypanosome;

Country	# of BSL-2/3/4[a]	Type of Lab[b]	Funder[c]	Status[d]	Capability
	C. burnetii, Ebola Zaire)		Tuberculosis Project, affiliated with University of Zambia's School of Medicine and the London School of Hygiene & Tropical Medicine; no mention of funding		Reference Lab/Culture Collection
Zimbabwe	1 BSL-3 No known BSAT 3 possible BSL-3 (no details)	National TB Reference Laboratory, Min Health	No funding described	Operational	National Reference Lab/Culture Collection, works with MDR/XDR TB, HIV

[a] BSL indicates BSL-2 is included if there is no known BSL-3.
[b] Names of lab(s) and affiliation (academic, commercial, agriculture, military, hospital)
[c] Donor paying for lab/operations, last date of funds; specialties (WHO center TB, Polio, flu, etc.)
[d] Status: operational, not operational, unknown.

NOTES: Although information is available on laboratory programs in several other countries that are not listed as low-income or lower- to middle-income countries by the World Bank, such as Jordan and Trinidad and Tobago, it was decided to limit the scope of this illustrative list.

Abbreviations:

BLRI, Bangladesh Livestock Research Institute
BSAT, Biological Select Agents and Toxins
CDIL, Central Disease Investigation Laboratory, Dhaka Bangladesh
CPHL, Central Public Health Laboratories
FMPOS, Faculty of Medicine, Pharmacy, and Odonto-Stomatology
GHESKIO, Institute of Infectious Diseases and Reproductive Health, Haitian Group for the Study of Kaposi's Sarcoma and Opportunistic Infections
icddr,b, International Centre for Diarrhoeal Disease Research, Bangladesh
IEDCR, Institute of Epidemiology, Disease Control & Research, Bangladesh
INRB, National Institute of Biomedical Research
IP, C, Institut Pasteur in Cambodia
IP, Cd'I, Institut Pasteur in Côte d'Ivoire
MDC, Microbial Depository Center
MIRCEN, Egypt Microbial Culture Collection, Cairo Microbiological Resources Centre
MVDL, Mandalay Veterinary Diagnostic Laboratory
NaVRI, National Veterinary Research Institute
PC,C, Pasteur Center, Cameroon
PCPHL, Palestinian Central Public Health Laboratory
UVRI, Uganda Virus Research Institute
YNRL, Yangon National Reference Laboratory